Kekulé Centennial

A symposium co-sponsored by
the Division of History of
Chemistry, the Division of
Organic Chemistry, and the
Division of Chemical Education
at the 150th Meeting of
the American Chemical Society,
Atlantic City, N. J.,
Sept. 15–16, 1965.

O. Theodor Benfey, *Symposium Chairman*

ADVANCES IN CHEMISTRY SERIES 61

AMERICAN CHEMICAL SOCIETY

WASHINGTON, D. C.   1966

Advances in Chemistry Series

Robert F. Gould, *Editor*

AMERICAN CHEMICAL SOCIETY PUBLICATIONS

195490

FOREWORD

ADVANCES IN CHEMISTRY SERIES was founded in 1949 by the American Chemical Society as an outlet for symposia and collections of data in special areas of topical interest that could not be accommodated in the Society's journals. It provides a medium for symposia that would otherwise be fragmented, their papers distributed among several journals or not published at all. Papers are refereed critically according to ACS editorial standards and receive the careful attention and processing characteristic of ACS publications. Papers published in ADVANCES IN CHEMISTRY SERIES are original contributions not published elsewhere in whole or major part and include reports of research as well as reviews since symposia may embrace both types of presentation.

CONTENTS

Kekulé stamps commemorating the 100th anniversary of the benzene formula. Issued during the centennial year by Germany and Belgium, the two countries in which he lived and worked.

PREFACE

Perhaps it is fitting to begin this volume the same way the Kekulé celebration was begun—with a quotation by the man we honor:

We all stand on the shoulders of our predecessors. Is it then surprising that we can see further than they? If we follow the roads built by our predecessors and effortlessly reach places they have attained only after overcoming countless obstacles, what special merit is it if we can penetrate further into the unknown?

Kekulé spoke these words at the 25th anniversary of the formula that he proposed 100 years ago. Today we honor Kekulé, the giant on whose shoulders we now stand. The occasion for this celebration is not the centennial of his birth or death but the centennial of a particular concept.

This is not the first benzene centennial celebration. Belgium has already celebrated Kekulé's benzene formula, using the occasion to announce the synthesis of some of the isomers of benzene. Coinciding with the American Chemical Society's Kekulé symposium, a Kekulé celebration was held in Bonn by the German Chemical Society. Kekulé had his vision of the benzene ring in Belgium, but for many years he taught in Bonn. The American Chemical Society was officially invited to send a delegate to the German celebration, and Professor Saul Winstein of U.C.L.A. was appointed to bring the greetings of the American Chemical Society to its sister society in Germany.

The Kekulé celebrations, both in Bonn and Atlantic City, had historical and contemporary facets. Papers were presented on historical subjects and on the current state of aromatic chemistry. The Atlantic City symposium was co-sponsored by the Divisions of History of Chemistry, Organic Chemistry, and Chemical Education. This volume presents only the historical papers.

Special articles commemorating the benzene formula have appeared in *Angewandte Chemie* [**77**, 770 (1965)]; *Chemical and Engineering News* [**43**, 90 (June 25, 1965)]; *Journal of Chemical Education* [**42**, 266 (1965)]; *Chemistry* [**38**, 6 (January 1965)].

It is our hope that this volume will contribute to the understanding by scientists of the conceptual history behind modern structural chemistry, the significance of the individual in the ongoing path of science, and the interaction of science, technology, and society.

Richmond, Ind.   O. Theodor Benfey
January 1966

Kekulé and the Architecture of Molecules

GEORGE E. HEIN

Harvard Medical School, Boston, Mass.

Kekulé's interest in architecture influenced his structural theory. His opinions are compared with those of his contemporaries—Frankland, Butlerov, Ladenburg, and Wislicenus. Evidence of Kekulé's concern for the spatial arrangement of atoms in molecules is obtained from his introspective comments, his use of tetrahedral models, and some arguments in support of his benzene structure. He did not explicitly anticipate the solutions proposed by van't Hoff and le Bel. Clarification of Kekulé's views is valuable in suggesting approaches to the teaching of organic structural theory. It is historically inaccurate and confusing to suggest that the Kekulé-Couper theory considered molecules as two-dimensional entities. The theory required no specific arrangement in space but did refer to chemical relations between atoms in three dimensions.

The theory of structural organic chemistry as developed in the 19th century may be the most fruitful conceptual scheme in all the history of science. Among western chemists, one name is particularly associated with the exposition of the theory: August Wilhelm Kekulé. We now recognize that others also made major contributions, and the claims of Couper and Butlerov especially have been justly advanced in recent years.

I would like to concentrate on Kekulé and discuss what appears to be an enigmatic situation. There are two incongruities in our understanding of Kekulé. One is that although he is considered the founder of a powerful and fantastically productive theory, Kekulé is essentially unknown except by chemists and professional historians. It is incorrect to attribute the development of organic structural theory exclusively to him, just as it is incorrect to assume that Darwin was the only naturalist who believed in natural selection or that Pasteur was the only experimenter who opposed spontaneous generation. Yet Darwin and Pasteur are universally recognized for their contributions while Kekulé is barely mentioned outside chemical circles. The publication of his paper (5) on chemical structure in 1858 is one of the great events of a few months in which, among other things, "The Origin of Species," "Das Kapital," and a key work by Pasteur on fermentation were also published.

When we consider that Kekulé was an intensely thoughtful man and that he delivered a major address on creativity and the psychological processes related to discovery (9), it is even more noteworthy that he is excluded from the mainstream of intellectual history.

The isolation of Kekulé may be attributed partly to chemists themselves. This brings us to the second enigma. Kekulé originally began to study architecture. He was dissuaded from this by Liebig's lectures at Giessen, but he retained an "architectural sense" all his life. It is correct to call him "the architect of organic chemistry." Now, an essential feature of anything we might call architecture is a concern for arrangement in space. Specifically, architecture deals with a three-dimensional, "full-bodied" view of the world and the components which structure it.

Many chemists, including the authors of many introductory texts, have ignored Kekulé's concern with space. The standard view (16) is that Kekulé developed a two-dimensional approach to organic chemistry and that his theory was later extended to three dimensions by van't Hoff and le Bel. I believe that this contention is incorrect. I shall try to demonstrate this error and to point out that besides misrepresenting Kekulé's position, this interpretation of his contribution has resulted in poor pedagogy.

Organic Chemistry and Atomic Theory around 1860

It might appear that a clear statement of Kekulé's views concerning the disposition of atoms in space would be an easy matter. Actually, this is a complex question which requires understanding the status of atomic and molecular theory at that time.

In the middle of the 19th century, the atomic theory was considered a good hypothesis, but it was not accepted unequivocally by most scientists, as is the case today. In particular, there was no compelling reason to equate the atoms and molecules which were used to explain the behavior of gases in terms of kinetic theory with the molecules of organic chemistry (15). At that time two kinds of molecules were discussed: the "physical" molecule and the "chemical" molecule. Every organic chemist who wrote about molecular structure around 1860 included a caveat to the effect that he was describing an apparent structure which might or might not be identical with the actual physical description of a molecule. The problem was how to develop a formal system which could explain the chemical transformations and cases of isomerism observed in organic chemistry. Kekulé's contribution was to develop a detailed formal system which pictured molecular structure in space and which accounted for a large number of observed phenomena. He was the only one who included three-dimensional spatial features as an integral part of his formal system—the only one who thoroughly considered the architecture of molecules.

Representative Views on Chemical Structure

A brief description of the positions taken by other chemists during this period may help to differentiate their views from Kekulé's.

William Frankland. Frankland contributed significantly to the concept of valency. He worked with the new ideas of molecular structure but treated them rather simply.

In a typical paper (*4*), published in 1867, Frankland and Duppa discuss the structure of hydroxy acids. The symbols they use (Figure 1) are clearly related to Crum Brown's two-dimensional models. They state in a footnote:

It is hardly necessary to repeat Crum Brown's comment that these formulas only represent the chemical and not the physical position of the atoms.

In the entire paper there is no statement which can be interpreted as referring to the spatial arrangement of the atoms. The lines which connect the croquet-ball atoms are lines of valence, not of structure.

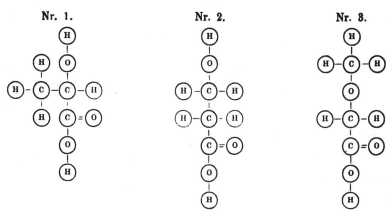

Oder symbolisch ausgedrückt :

Nr. 1. Nr. 2. Nr. 3.

$\begin{cases} \text{CMeHH o} \\ \text{COH o} \end{cases}$ · $\begin{cases} \text{CH}_2\text{H o} \\ \text{CH}_2(\text{COH o}) \end{cases}$ oder $\begin{cases} \text{CH}_2\text{H o} \\ \text{CH}_2 \\ \text{COH o} \end{cases}$ · $\begin{cases} \text{CH}_2\text{Me o} \\ \text{COH o} \end{cases}$

Figure 1.   Crum Brown's flat, "croquet-ball" formulas as used by Frankland to represent the isomeric acids $C_3H_6O_3$.

Butlerov. If there is anyone who can challenge Kekulé as originator of the concepts of structural chemistry, that person is certainly A. M. Butlerov. The term "molecular structure" is his, as is the concept that for every compound there is only one structure, and for every symbolic structure there is only one compound. But Butlerov also did not consider the arrangement of atoms in space. He carefully limited his remarks to chemical structure as opposed to physical structure, and he opposed the use of the term "topography of atoms" instead of "molecular structure" "because it involves the concept of the position of atoms in space." He wished to keep chemical relations separate from both the questions of the actual existence of atoms and of their positions in space.

Butlerov entered into polemical arguments with his critics concerning the problems of molecular structure. In the same paper in which he expresses his dissatisfaction with the term "topography of atoms," he also expresses the view that someday the position of atoms in space will be established. However, he clearly considers this a problem for the future, and he presents his own arguments in a nonspatial framework.

One of Butlerov's most powerful contributions to the concepts of molecular structure was his insistence on the one-to-one correspondence between compounds and the structure used to represent them. He argued, as we do today, that a successful structural theory would allow one, and only one, structure to be written for each compound. He criticized (2) Kekulé for an apparent contradiction. In the first volume of his text, Kekulé claims that he is only discussing chemical transformations and that different structures may be assigned to the same compound, but he actually goes on to discuss the structure of the compounds in terms of the relative relations of the atoms to each other, independent of any chemical transformation.

Butlerov's criticism is correct; Kekulé did attribute more "structure" to his formulas than he claimed. As the years went by and he developed the system, Kekulé's concern with structure increased. Specifically, he was concerned with chemical structure in space. The same is not true of Butlerov. The latter does use the concept of matching structures, and he empirically discovered isomers with much more sophistication than did Kekulé. However, the structures drawn have formal significance not spatial significance. In a discussion of isomerism among butanes and butenes (3), he correctly stated that only two isomeric butanes are possible, but he considered no less than nine theoretically possible butenes (Figure 2). In a formal sense this was correct, but, clear and neat as the argument is, one recognizes that he is just not concerned with translating his system into a three-dimensional model.

Ladenburg. Another formal use of structural theory is provided by W. Ladenburg's famous prism formula for benzene. It is absolutely

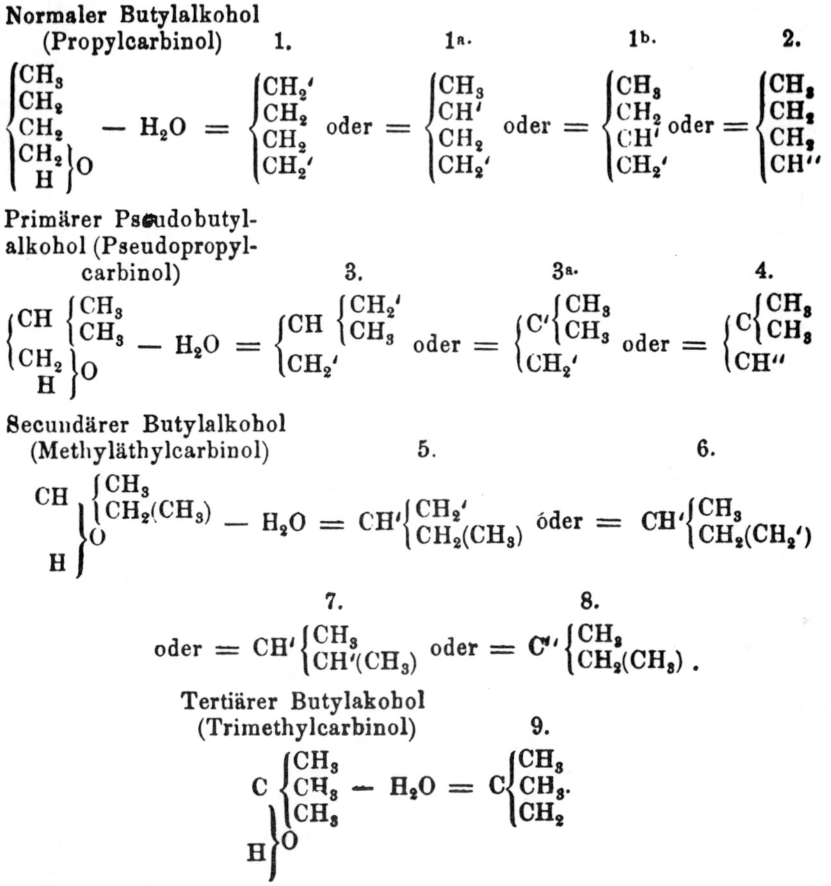

Normaler Butylalkohol
(Propylcarbinol)   1.   1ᵃ·   1ᵇ·   2.

$$\begin{Bmatrix}CH_3\\CH_2\\CH_2\\CH_2\\H\end{Bmatrix}O \; - \; H_2O \; = \; \begin{Bmatrix}CH_2{}'\\CH_2\\CH_2\\CH_2{}'\end{Bmatrix}\text{oder} = \begin{Bmatrix}CH_3\\CH'\\CH_2\\CH_2{}'\end{Bmatrix}\text{oder} = \begin{Bmatrix}CH_3\\CH_2\\CH'\\CH_2{}'\end{Bmatrix}\text{oder} = \begin{Bmatrix}CH_3\\CH_2\\CH_2\\CH''\end{Bmatrix}$$

Primärer Pseudobutyl-
alkohol (Pseudopropyl-
        carbinol)   3.   3ᵃ·   4.

$$\begin{Bmatrix}CH\begin{Bmatrix}CH_3\\CH_3\end{Bmatrix}\\CH_2\\H\end{Bmatrix}O \; - \; H_2O \; = \; \begin{Bmatrix}CH\begin{Bmatrix}CH_2{}'\\CH_3\end{Bmatrix}\\CH_2{}'\end{Bmatrix}\text{oder} = \begin{Bmatrix}C'\begin{Bmatrix}CH_3\\CH_3\end{Bmatrix}\\CH_2{}'\end{Bmatrix}\text{oder} = \begin{Bmatrix}C\begin{Bmatrix}CH_3\\CH_3\end{Bmatrix}\\CH''\end{Bmatrix}$$

Secundärer Butylalkohol
(Methyläthylcarbinol)   5.   6.

$$CH\begin{Bmatrix}CH_3\\CH_2(CH_3)\\O\\H\end{Bmatrix} \; - \; H_2O \; = \; CH'\begin{Bmatrix}CH_2{}'\\CH_2(CH_3)\end{Bmatrix} \; \text{óder} = \; CH'\begin{Bmatrix}CH_3\\CH_2(CH_2{}')\end{Bmatrix}$$

7.   8.

$$\text{oder} = \; CH'\begin{Bmatrix}CH_3\\CH'(CH_3)\end{Bmatrix} \; \text{oder} = \; C''\begin{Bmatrix}CH_3\\CH_2(CH_3)\end{Bmatrix}.$$

Tertiärer Butylakohol
(Trimethylcarbinol)   9.

$$C\begin{Bmatrix}CH_3\\CH_3\\CH_3\\O\\H\end{Bmatrix} \; - \; H_2O \; = \; C\begin{Bmatrix}CH_3\\CH_3\\CH_2\end{Bmatrix}.$$

Figure 2. Butlerov's analysis of the isomeric butenes. Each half unsaturation is indicated by '. The inclusion of nine isomers is formally correct but requires consideration of structures in which both unsaturations are on the same atom as well as structures in which the unsaturations are on nonadjacent atoms. Butlerov had no criterion for rejecting any of these structures.

impossible to appreciate Ladenburg's position unless we realize that he was treating structural theory as a purely formal system with no implications concerning the relative positions of atoms in space. In the paper which first introduced the prism formula (*10*), he considered three possible structures which all met what he considered the necessary requirements. Among these requirements were (a) that all six hydrogens be identical, and (b) that there be two pairs of equivalent positions in disubstituted derivatives. He added:

If, as is common, graphic formulas are used to visualize constitution, then the geometric relations determine the relations between the atoms. By means of a figure we do not presume to indicate the spatial position of the atoms.

Ladenburg used a model which translated molecular properties into geometric terms. Ladenburg himself recognized the limitations of the prism formula. He admitted (11) in 1875 that his formula predicted incorrectly that the three ring hydrogens in mesitylene should be non-equivalent, but as late as 1887 he still insisted (12) that the prism formula "is the only one which presents a clear and complete picture of isomerism in the benzene series." He is correct in the sense that the geometric relations between the apexes of a prism can be taken as a model for the relative positions of substituents in benzene as long as no correlation exists between the geometric space of the model and the space occupied by atoms in a molecule.

Wislicenus. Another who was obviously concerned with the arrangement of atoms in space before van't Hoff and le Bel was J. A. Wislicenus. His work on isomeric hydroxypropionic acids revealed more isomers than could be accounted for by the current structural theory. In particular, he could not explain the three lactic acids—two optically active ones derived from biological tissues, and an inactive, synthetic form. He concluded (17):

Since structural formulas only represent the manner in which atoms are connected, we must admit that if two different substances have the same structural formulas, their differing properties must result from differences in the spatial arrangements of the atoms within the molecule.

Wislicenus did not follow up these comments with any specific model. Perhaps it was because he so clearly recognized the type of answer that was needed, without having committed himself to any particular solution, that he rapidly became van't Hoff's champion and introduced the latter's work into Germany.

The brief summaries above represent typical positions held by chemists in the period 1860–1870. Frankland considered structural theory primarily as a device for easy symbolization; Butlerov used it to predict compounds, Ladenburg tried to develop pure geometric relations, and Wislicenus struggled with the poorly defined spatial relations.

Kekulé's Views

Kekulé's approach to organic structural chemistry was somewhat different from the positions outlined above. I do not wish to claim that it was better, but I do assert that it incorporated spatial arrangements as an integral part of a formal system. Kekulé was willing to sacrifice rigor in order to tie his concepts to a reasonable three-dimensional model. In the benzene controversy and in many other disputes, Kekulé's view prevailed partly because it permitted chemical structures to be visualized in common, three-dimensional space.

Benzene Structure. Kekulé proposed (6) his famous benzene formula in 1865. Four years later Ladenburg criticized this and suggested alternative structures as discussed above. Kekulé's answer (8) to this attack is characteristic of his approach.

We must assume that the atoms of a polyatomic molecule are arranged in space so that all the attractive forces are satisfied.

The thoughts expressed by the scheme above corresponded to the assumption of an arrangement of the atoms in one plane. The model which I recommended some time ago, in order that we may visualize the linkages of atoms, leads to a figure in which all the atoms are arranged in one plane.

Kekulé does not argue in support of his benzene *formula;* he argues in support of his views concerning the structure of the benzene *molecule.* Essentially similar views were expressed in earlier publications on benzene. Kekulé visualized the molecule in space—he pictured the atoms arranged in a plane in space.

The insistence on a meaningful model which corresponds to a reasonable arrangement of the atoms explains why Kekulé was willing to stick with his hexagon formula even when he had to write two structures for one molecule (8). This drastic step can hardly be justified any other way. The hexagon formula explains the isomer relations in the benzene series, and it makes sense out of these in space. It gives a simple picture for the formation of mesitylene from benzene (Figure 3), and it explains anhydride formation from phthalic acid. These and other observations are discussed by Kekulé with reference to the spatial arrangement of the atoms.

Butlerov and Unsaturation. Kekulé's views on the structure of benzene were certainly inconsistent with his earlier statements about our inability to determine the arrangement of atoms in space. Butlerov's criticism of 1863 was correct and partly prophetic. Kekulé increasingly discussed the spatial arrangement of atoms and not just their valency relations. This shift in position was not difficult for Kekulé because he was concerned with the chemical and not the physical atom. His was a formal scheme but one which was developed in space.

Kekulé considered the double bond as an unsaturation between two atoms. Consistent with the development of his system in space, he visualized this double bond as a property shared by adjacent atoms. This is in contrast to Butlerov's formal treatment of unsaturation as indicated in Figure 2. Perhaps at the time there was little experimental justification for assuming that unsaturation must exist between adjacent atoms. I suspect that Kekulé reached this conclusion partly because it was the most reasonable one in terms of his specific spatial model.

Molecular Models. One of the most powerful arguments in support of Kekulé's concern with space can be based on the type of models he

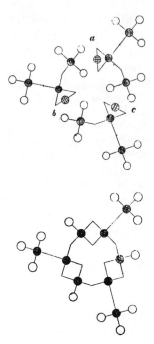

Figure 3. The formation of mesitylene (lower drawing) from three molecules of acetone (upper drawing) as depicted by Kekulé. The formation of the product from the reactants is clearly visualized in a spatial manner.

used. In the 1860's it first became popular to employ models in order to understand structural relations, and a number of different types, were developed.

The only ones which seem to have portrayed the valences of a single atom as directed in space were those developed by Kekulé. In ·fact, he used a tetrahedral model for his carbon atom (7)!

As usual, the model was not supposed to represent necessarily the physical atom, but Kekulé did consider it the best possible model of the chemical atom. He criticized (7) Crum Brown's models:

It has other drawbacks. It only appears to fill space, where as it actually places all the atoms in one plane. The model does not add anything to a drawing.

Kekulé especially recommends his tetrahedral model because it allows a simple visualization of double and triple bonds (Figure 4).

The implications of Kekulé's use of a tetrahedral model are obvious. For example, he employed them in his lectures, which van't Hoff attended in the early 1860's (*1*). However, the important point is not, I think, that Kekulé deserves credit for development of the concept of the tetrahedral atom. The important point is that Kekulé was discussing molecular structure in space.

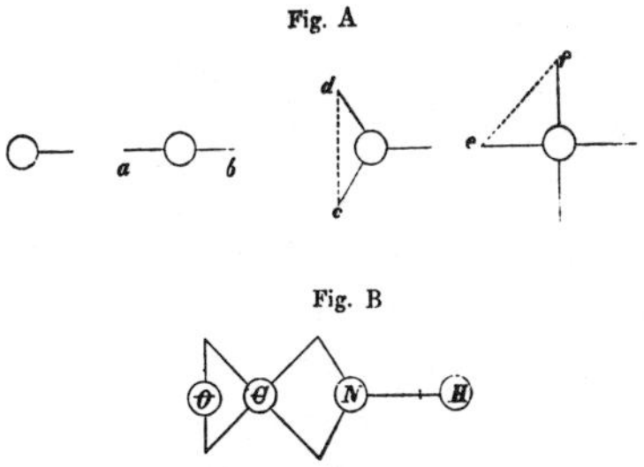

Figure 4. *Kekulé's description of his tetrahedral model for carbon compounds. His original text reads:*
"It is then possible to combine the atoms not only by means of one, but also of two valences [Figure B]. This method is sufficient for most of the common cases, but it is still very limiting. It cannot be used to represent the combination of three valences of carbon with one other atom of carbon or nitrogen.
Even this difficulty can be overcome, at least in a model, if the four valences of carbon, instead of being projected in a plane, are directed in the direction of hexagonal axes from the atom so that they form a tetrahedron. In doing this, the lengths of the wires which represent the valances are chosen so that their lengths are equal. Simple directions, whose description could be out of place here, permit the wires to be joined linearly or in any desired angle.
A model of this type permits the representation 1, 2 or 3 valence relationships, and it seems to me that it does everything that a model could conceivably do."

Background of Kekule's Views

Victor Mayer. Besides the direct evidence of Kekulé's writings to support the view that he was concerned with the arrangement of atoms in space, we have the indirect comments of his contemporaries and students. One who was deeply involved in the development of structural chemistry was Victor Mayer. As Kekulé's student he coined the word "stereochemistry." He published a series of lectures on stereochemistry in 1890, which includes a good deal of historical material (*14*). He asserts that the tetrahedral model was used before van't Hoff and le Bel, and he previously (*13*) refers to it as the Kekulé-van't Hoff model. Without claiming priority for himself, he states that he had used the model since 1871 and demonstrates how a spatial model is necessary to explain the number of isomers known at that time in the series CH_3Cl, CH_2Cl_2, $CHCl_3$, etc.

Comments by German scientists about ideas attributed to non-German scientists must be taken with a grain of salt, but Mayer's tone is clear; the view that the structural chemical models represented a three-dimensional picture existed before 1874. A specific model—the tetrahedral one—was available and was well known, especially to anyone who had attended Kekulé's lectures.

Kekulé's Dreams. If no other evidence were available, Kekulé's own words concerning the origin of the structural theory and the concept of the benzene ring would convince us that his thoughts concerned the arrangement of atoms in space. He described (9) his reveries at length at a festival to honor the 25th anniversary of his paper on benzene structure. The structural theory originated during a daydream in which:

The atoms were gamboling before my eyes. . . I saw how, frequently, two smaller atoms united to form a pair; how a larger one embraced two smaller ones; how still larger ones kept hold of three or four of the smaller; whilst the whole kept whirling in a giddy dance.

And later,

Again the atoms were gamboling before my eyes. . . My mental eye rendered more acute by repeated visions of this kind, could now distinguish larger structures of manifold conformation; long rows, sometimes more closely fitting together, all twining and twisting in snake-like motion. But look! What was that? One of the snakes had seized hold of his own tail, and the form whirled mockingly before my eyes.

Kekulé was neither so naive nor (perhaps) so visionary that he believed that his reveries had revealed to him the actual structure of atoms. But these remarks do reveal the kind of approach he had to chemical structure. Whatever symbols Kekulé might choose to portray his theory, these symbols were only an attempt to represent a model which clearly existed in space. The model might have only a tenuous connection with the physical reality of atoms, but the model existed in space.

This view is in clear contrast to those presented before. It differs sharply from Ladenburg's concern with geometric relations or from Butlerov's formal treatment of chemical formulas. In some ways the more formal treatment by Butlerov was necessary to clarify the concepts of molecular structure and to provide for the powerful predictive value of structural formulas. Nevertheless, Kekulé's imaginative architectural approach is the direct predecessor of van't Hoff's and le Bel's synthesis of the chemical and the physical atom.

van't Hoff and le Bel. If we attribute the general concept of three-dimensional structure and even the specific concept of a tetrahedral carbon atom to Kekulé, then what contribution did van't Hoff and le Bel make to organic chemistry? They provided an explanation for one type of isomerism by pointing out particular consequences of three-dimensional models. van't Hoff discussed his theory specifically in terms of a tetrahedral model while le Bel spoke more generally in terms of any one of a class of space-filling models. They noted the asymmetry possible in tetrasubstituted carbon compounds, and they explained observed optical activity in terms of their models.

However, this contribution, significant as it may be, was not their chief claim to fame. It was not this which aroused the ire of Kolbe. Much more important was their synthesis of the "chemical" and the "physical" atom. van't Hoff and le Bel claimed that the fanciful structural theory of the organic chemist not only was useful, not only could explain the facts, but that it also happened to be "true."

They explained a physical property of molecules—their optical rotary power—and therefore one that was related to their physical structure. The explanation they offered was to use the space-filling models of the organic chemist. Since, in the 19th century, everyone believed that physical measurements revealed the true nature of matter, van't Hoff's and le Bel's correlation of isomerism with a physical property provided a foundation for organic structure theory. They did not introduce the third dimension into organic chemistry, but they did legitimize the spatial arrangement of atoms in molecules.

The Significance of Kekulé's Structural Theory

Various factors have combined to foster the common, incorrect view that Kekulé developed a two-dimensional approach to organic chemistry. It is the task of the historian to attempt to set the record straight. However, if this were our only concern, it would represent a sterile effort.

Much more significant than the question of what Kekulé's theory was really like is the question of how organic structural theory is and should be taught. The historical misconception concerning the origins of the theory have been reflected in a parallel pedagogic approach. For many years organic texts presented first an essentially two-dimensional statement of organic structural theory and did not introduce a spatial concept until much later to explain optical activity. In the interim, students remained puzzled about the identity, or lack of it, for many compounds when projected in various ways on the plane of the text or the blackboard.

Even current texts, although they usually contain a statement in an early chapter concerning the tetrahedral carbon atom, often fail to stress the fact that all formulas written on a page are only two-dimensional representations of a three-dimensional model. The extension into three-dimensional space appears to me to be an integral part of Kekulé's theory and should be presented as such.

In recent years increased concern with such topics as conformational analysis or stereochemical evidence for reaction mechanisms has forced a concern with stereochemistry into earlier sections of the chemistry curriculum. Some authors apparently feel that the tetrahedral model cannot be justified in terms of classical organic chemistry, and therefore they substitute a nonhistorical approach. The spatial arrangement of molecules is presented as a consequence of modern bonding theories. This approach can be highly successful if it is thorough and includes a considerable amount of material concerning the electronic properties of atoms. Unfortunately, most texts do not contain enough background material along this line nor can they be understood by the majority of students who are enrolled in organic chemistry courses.

I maintain that structural organic chemistry can be profitably taught on the basis of the early concepts. The most significant theory is that of Kekulé. By presenting it fully, a satisfying and complete approach to organic structures can be developed. In addition to Kekulé's views, Butlerov's concept of the relation between formulas and structure and van't Hoff's and le Bel's clear definition of the properties of the three-dimensional atom make for a solid theory which is still serving the organic chemist admirably.

Literature Cited

(1) Anschutz, R., "August Kekulé," Vol. 1, Verlag Chemie, Berlin, 1929.
(2) Butlerov, A. M., Zeit. Chemie **6**, 500 (1863).
(3) Butlerov, A. M., Ann. **144**, 1 (1867).
(4) Frankland, E., Duppa, B. F., Ann. **142**, 1 (1867).
(5) Kekulé, A., Ann. **106**, 129 (1858).
(6) Kekulé, A., Ann. **137**, 129 (1865).
(7) Kekulé, A., Zeit. Chemie **3**, 214 (1867).
(8) Kekulé, A., Ann. **162**, 77 (1872).
(9) Kekulé, A., Ber. **23**, 1302 (1890).
(10) Ladenburg, A., Ber. **2**, 140 (1869).
(11) Ladenburg, A., Ann. **179**, 163 (1875).
(12) Ladenburg, A., Ber. **20**, 62 (1887).
(13) Meyer, V., Ann. **180**, 192 (1875).
(14) Meyer, V., "Ergebnisse und Ziele der Stereochemischem Forschung," Heidelberg, 1890.
(15) Metz, J. T., "European Thought in the Nineteenth Century," Vol. 1, p. 305, Blackwood and Sons, Edinburgh, 1914.
(16) Singer, C., "A Short History of Scientific Ideas to 1900," p. 485, Oxford, 1959, gives a good example of this view.
(17) Wislicenus, J., Ann. **166**, 47 (1873).

RECEIVED January 3, 1966.

Kekulé, Butlerov, Markovnikov: Controversies on Chemical Structure from 1860 to 1870

HENRY M. LEICESTER

College of Physicians and Surgeons, University of the Pacific, San Francisco, Calif.

V. V. Markovnikov was a student of Butlerov and an ardent supporter of his master's claim to have originated the structural theory of organic chemistry. More aggressive than Butlerov, he took a leading part in the polemics against Kekulé and his supporters, which enlivened the decade from 1860 to 1870. At the same time he devoted himself to experimental studies on the reciprocal effects of atoms and groups in organic compounds and their effect on reaction mechanisms. His well known "rule" was one of the results of these studies, which foreshadowed the later development of the electron theory of organic reactions.

Four major developments in the latter part of the nineteenth century were responsible for making the chemistry of carbon compounds comprehensible and supplying the foundation upon which organic chemistry has since developed. These were: (1) the recognition of the tetravalency of carbon and its ability to form chains with itself, (2) the structural theory which followed directly from this and which showed that the properties of carbon compounds depended on the arrangement of the atoms in the molecule, (3) the recognition that the various groups in the molecule affected each other reciprocally, and (4) the realization that these effects were three dimensional and that spatial factors had to be considered.

Kekulé

Kekulé and Couper recognized the first of these developments in 1858, and their priority and originality have never been questioned. The stereochemical viewpoint, though suggested in the statements of a number of chemists, had to await the definitive work of van't Hoff and Le Bel in 1874 before it became a usable part of organic chemistry.

The period between these dates, essentially the years from 1860 to 1870, is that in which the modern structural theory came of age. It is divided midway by the benzene ring theory of Kekulé, but this theory is really a part of the whole development. It is a period full of controversies, claims to priority, and many polemical papers. An account of some of these controversies shows how difficult it was to discard the old, even when all the essentials for establishing the new were at hand.

The two figures around which the major controversies raged were Kekulé himself and Alexander Mikhailovich Butlerov of the University of Kazan in Russia (14). The two principals did not take a very active part in the battle, but their supporters were extremely vocal. Both personal and nationalistic prejudices entered the picture. Many prominent German chemists supported the claims of Kekulé while Butlerov's chief support came from his student, Vladimir Vasil'evich Markovnikov.

After his paper of 1858 Kekulé certainly did not abandon type formulas. He himself had added the marsh gas type to the older types of Gerhardt, and he seemed indisposed to give this up and to follow the logical consequences of his own new theory. In the first volume of his influential textbook, circulated in parts from 1859 and published in full at Erlangen in 1861, he was very specific (12):

From the considerations previously given on the behavior of chemical metamorphoses, the alterations of radicals, etc., it is now evident that for most substances, different rational formulas are possible and that in many cases one rational formula cannot show all the metamorphoses equally (e. g., in hydrocyanic acid, etc.); from this it follows that one and the same substance can be expressed by different chemists in different rational formulas.

If we write acetic acid and propionic acid in the usual way, as

$$\text{Acetic acid} \qquad\qquad \text{Propionic acid}$$

$$\left.\begin{array}{l}\mathrm{C_3H_2O} \\ \mathrm{H}\end{array}\right\}\Theta \qquad\qquad \left.\begin{array}{l}\mathrm{C_3H_5O} \\ \mathrm{H}\end{array}\right\}\Theta$$

we allow these formulas to express a great number of metamorphoses—namely, all the metamorphoses in which methyl compounds are obtained from acetic acid and ethyl compounds from propionic acid. In order to express these changes, a number of formulas have been proposed in which methyl $\mathrm{CH_3}$ or ethyl $\mathrm{C_2H_5}$ are contained as radicals. The notable recent discovery of Wanklyn of the formation of propionic acid by the action of carbon dioxide on sodium ethyl shows plainly that along with the methyl and ethyl radicals in both acids the radical of carbonic acid must be assumed; we thus obtain the rational formulas

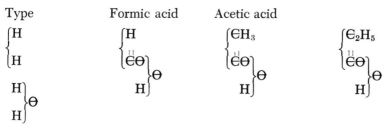

Type   Formic acid   Acetic acid

according to which they can be considered as belonging to the type $H_2 + H_2\Theta$ . . . . .

Which of the various rational formulas should be used for a given case depends on the purpose. The right to assume different rational formulas for the same substance cannot be doubted in view of the considerations given. Therefore we must naturally keep in mind that the rational formulas are only transformation formulas and not constitutional formulas; that they are nothing else than expressions of the metamorphoses of the compound, and comparisons of different substances with each other, and in no way express the constitution, that is, the position of atoms in these compounds.

This should be especially emphasized because, strange to say, some chemists are still of the view that from a study of chemical metamorphoses one can derive the constitution of compounds with certainty and can express in a chemical formula this position of the atoms. That this last is not possible does not need special proof; it is self evident that one cannot show the position of atoms in space, even if one had investigated this, on the plane of the paper by putting letters together; for this one would need at least a perspective drawing or a model. But it is likewise clear that one cannot determine the position of atoms in a specific compound by a study of metamorphoses, because the way in which the atoms leave a changing and decomposing compound cannot indicate how they are arranged in the existing and unaltered compound. Certainly it must be considered a problem for research workers to discover the constitution of the materials, and thus, if you will, the position of the atoms, but this can certainly not be accomplished by the study of chemical changes, but only by comparative studies of the physical properties of the unchanged compounds. It will then perhaps be possible to draw up constitutional formulas of chemical compounds which naturally must then remain the same for one and the same compound. But even when this is successful, the various rational formulas (transformation formulas) will always be needed because it is evident that with atoms arranged in a given way, a molecule will split in different ways and thus can give fragments of different size and different composition (*13*).

Butlerov

At about the same time Butlerov at Kazan was developing his ideas on chemical structure. He also was reluctant to discard the type theory, but after his visit to western Europe in 1857–1858 when he encountered the ideas of Kekulé and Couper, he began to consider possible modifications of the older views. Even by 1860, however, he had not abandoned them. This is shown in the dissertation of his student, Markovnikov (*15*).

In the autumn of 1860 Markovnikov presented a thesis for the degree of Candidate in the Juridical Faculty of the University of Kazan entitled "Aldehydes and Their Relation to Alcohols and Ketones" (*2*).

Although Markovnikov had originally intended to specialize in economics and hence had enrolled in the Finance Division of the Juridical Faculty, he had been attracted to chemistry by the teaching of Butlerov, and thus his dissertation undoubtedly reflects the latter's ideas. Moreover, the copy of the dissertation in the library of the University of Kazan has penciled annotations by Butlerov in the margin which indicate that he considered it an excellent piece of work. In this dissertation Markovnikov wrote the formula for aldehydes as $C_nH_{n-1}O_2 \brace H$ O, a type formula, and then went on:

Accepting along with others the above formula for aldehydes, we do not intend to express by it the true constitution of the substance, but only those directions in which the organic part splits in the majority of double decompositions. At the present time all of our rational formulas can serve merely to show those changes which a given substance can undergo under the action of different reagents or due to which they may originate, and the more any formula can express these changes, the more rational it is.

Thus, in 1860, Butlerov had not yet made the final step toward his theory of chemical structure.

However, in the following year he had come to the modern view of the constitution of chemical compounds and the formulas by which to express it. He summed up his views in the term "chemical structure," a term which he was the first to use in the modern sense, though it had been loosely used by a number of Russian chemists in the previous decade (*3*).

On September 19, 1861 he presented his classical paper at the Congress of German Naturalists and Physicians in Speyer, "On the Chemical Structure of Substances" (*1, 8*). The key sentence in this paper was:

The well known rule which says that the nature of compound molecules depends on the nature, the quantity, and the arrangement of its elementary constituents can for the present be changed as follows: the chemical nature of a compound molecule depends on the nature and quantity of its elementary constituents and on its chemical structure.

In other words, as he constantly emphasized later, for each compound there can be only one formula, which denotes its chemical structure, and to each formula there must correspond only one compound.

This was the great break with the type theory and its rational formulas which could vary indefinitely for a single compound. As the polemical literature which follows shows, many Germans did not appreciate this and felt that Butlerov was merely juggling letters in an unnecessary manner. Butlerov also felt that there was a nationalistic element in these controversies, for in his report to the University of Kazan on his return to Russia in 1861 he wrote:

One feature of these German congresses is particularly striking to us foreigners, so strange that I cannot pass over it in silence; it is an aspiration to assert their own nationality at every opportunity. . . And there can be no doubt that this hypertrophy of the national feeling does plenty of harm to the Germans; it keeps them from duly recognizing every alien nationality (9).

Butlerov had a much more realistic concept of chemical constitution. Thus, in a reply to a letter from his friend Adolphe Wurtz complaining that structural formulas were more complicated than type formulas, he wrote in 1864:

I believe that those type formulas which suffice for most of the relations of bodies serve only to express their chemical structure, or, at least, the most important part of their structure. I believe that in pursuing the idea of the atomicity of elements we must express their structure whenever the body has been studied sufficiently. (4)

Butlerov continued to develop his theory and to apply it to a number of cases of isomerism. In the laboratory he was able with its aid to synthesize the first tertiary alcohol, *tert*-butyl alcohol. In 1864 he published the first textbook utilizing the concept of chemical structure completely, his "Introduction to the Full Study of Organic Chemistry." The German translation appeared in 1868, although by that time most chemists were using structural formulas in the modern way, usually without giving much credit to Butlerov. This was particularly galling to the Russian group because Kekulé in the second volume of his textbook, published in 1866 though circulated in parts from 1863, had largely abandoned type formulas without in any way indicating that this was not entirely his own idea. Nevertheless, Butlerov refused to be drawn into controversy except on one occasion when, in 1868, he felt himself forced to reply to a criticism by Lothar Meyer (17).

However, his claims were amply emphasized by Markovnikov, and it is to his papers that we must look for major statements of the priority of Butlerov.

Markovnikov

Although Butlerov and Markovnikov remained close friends throughout their lives, they were men of quite different characters, and this goes far to explain their differing responses to criticisms. Butlerov was always pleasant and diplomatic. He steered his way through the frequent quarrels within the faculty at Kazan between the progressive and the reactionary political factions as well as through the student uprisings against the faculty which were frequent at Kazan, as they tended to be at most Russian universities of the period, and he managed to retain the respect of both sides. This is shown by the fact that both at Kazan and later at St. Petersburg he was able to fight for his principles without losing his university position. He was also able to attain election to the Academy of Sciences, an institution which was dominated by the reactionary German party. Neither Markovnikov, nor even Mendeleev, were able to accomplish these feats. After he had left Kazan, it is reported that when difficulties arose, men would say, "If Butlerov had been here, this would not have happened."

Markovnikov, on the other hand, was an enthusiast and a born controversialist. Even in his Candidate dissertation he had made a number of sweeping generalizations, which Butlerov had indicated should be modified. He was ever ready to defend vigorously, if not always politely, whatever he believed. As a result of his quarrels with the less progressive members of the Kazan faculty, and later at the University of Moscow, he was forced to resign his professorships in these universities. He recognized these traits in himself, and refers to them often in his letters to Butlerov. Thus, in a letter to his teacher written from Germany in 1866 he said, "You yourself have always called me a fighter, and you know that I cannot cold-bloodedly endure anything in which I see error" (23).

Later, when he had succeeded Butlerov at Kazan and was engaged in a struggle with political opponents, he wrote to Butlerov, "Among us there is either complete indifference or intrigue. I regret very much that I cannot copy your coolness and optimism." Again he wrote, "All this filth and these quarrels sicken me to such an extent that I wish to avoid any sort of meeting. You will notice that they sicken *me*, and you have considered me to be a great wrangler" (24).

Theory of Chemical Structure

It was this enthusiastic and strongly partisan student who took up the cudgels and fought for the reputation of his teacher. The immediate cause of his first polemic was an article by Wilhelm Heinrich Heintz (1817–1880), professor at Halle, on "Ethyl Glycol Amide and Some Compounds of Ethyl Glycocol" (7). In this, Heintz stated that he was far from wishing to indicate the actual position of the atoms in a compound by his formulas and then went on to say:

The words "chemical structure" mean to me only the condition of chemical compounds in which the contained elements are bound to each other by different degrees of firmness. . . . Chemical structure is the condition of chemical compounds which they attain from the relative separation of their atoms from each other. We are indeed far from being able to measure or calculate these separations and we thus have no idea of the real positions of the atoms in compounds. However, the different degrees of attraction by which the atoms are joined together gives us at least the possibility of forming an approximate picture of the positions of the atoms. We can therefore properly speak of the structure of chemical compounds without, indeed, having a detailed knowledge of it, just as we can speak of electricity, magnetism, etc., without complete knowledge of these.

I know very well that these relative positions of the atoms in chemical compounds have already been considered much earlier, namely by Kekulé. Butlerov's service is in having first used the name chemical structure. This service is indeed not very important, merely to have found the best terminology. But after all, this does not matter. It would indeed be a greater service if a scholar of recognized reputation would give up a method of notation for which he has fought, but which has been bypassed, and would go on to better matters.

This was too much for Markovnikov. He was engaged in writing his dissertation for the Master's degree in Chemistry, "On the Isomerism of Organic Compounds," and he inserted into this a long defense of Butlerov and an attack, not only on Heintz, but also on Kekulé himself (*19*).

Since he knew that this dissertation would not be read outside of Russia, he wrote a long paper in which he translated into German a considerable part of the historical portion of his thesis, in which he had attacked Heintz and Kekulé. He submitted this to the *Zeitschrift für Chemie*, a journal which had been edited by Emil Erlenmeyer (1825–1909) (*6*) for several years. Erlenmeyer had succeeded Kekulé in teaching organic chemistry at Heidelberg when the latter moved to Ghent and had been especially friendly with Russian students, a number of whom had been attracted to Heidelberg by his interest. In recognition, the Russian government awarded him the Order of St. Anne in 1865. He had opened the pages of the *Zeitschrift für Chemie* to Russian chemists, and many of Butlerov's important papers had appeared in its pages, as well as a number of papers of a rather controversial character. As a result, it was said that the journal was read more widely in Russia than in Germany, and at the end of Erlenmeyer's editorship there were only 150 subscribers (*5*).

In 1865 the editorship was taken over by a board of three men, Fittig, Hübner, and Beilstein, all then at Göttingen. Friedrich Beilstein was a Russian of German descent and a strong member of the German party in Russian academic circles. This was, in general, the reactionary party with which the leading Russian chemists, including Butlerov, Markovnikov, and Mendeleev, were at odds. Beilstein was given particular charge of papers from Russia for the *Zeitschrift,* and he particularly desired to reduce the number of polemical papers which the journal published. It was to him that Markovnikov's defense of Butlerov came. He at once wrote Butlerov concerning it:

I have read the Markovnikov communication, and I find that it contains much that is true, but it does not present a new viewpoint and so runs counter to the tendency of our journal. However, since you are naturally interested in the paper, Herr M. has rightly remarked, and I am of the opinion, that it is a good thing for people to get the pure wine, and so the paper will be printed.

Nevertheless, Beilstein edited the paper so heavily that he afterwards apologized to Markovnikov. He later wrote to Butlerov:

Herr M. up to now has received very little thanks for his paper. In Berlin Baeyer has made the strongest protest, which on a visit he asked me to publish, and this I have done. It seems that it is the form of the paper which has offended many people (*10*).

The paper appeared under the title "On the History of the Theory of Chemical Structure" (*11, 18*). In it Markovnikov used his reply to the criticisms of Heintz to introduce his attack on Kekulé. He wrote:

Anyone who has carefully read the theoretical papers of Butlerov can scarcely doubt that his true services consist in the fact that he knew how to use the principle of chemical structure consistently, and always was anxious to follow strictly the consequence which stemmed from it. This permitted Butlerov to recognize the inconsistencies and contradictions in other theoretical ideas. I will not reproach Heintz along with Kekulé that he has not accepted the expression "chemical structure," but Kekulé has used this expression in a peculiar way which seems to me the more astonishing in that some of the ideas of Butlerov on chemical structure seem to me to be shared by Kekulé also. In his paper on "Different Methods of Explaining Isomerism" [1863] Butlerov tried to show specifically the unsuitability of types, especially the mixed types, and the similarity between the views of Kolbe and those of Kekulé. Now we see later, in the second part of volume 2 of Kekulé's textbook that no more mixed types appear, and, without a mention of Butlerov, Kekulé here speaks of the formulas of Kolbe in almost the same way as Butlerov had done.

The paper continues at some length to criticize Kekulé sharply, though not in the detail given in his Russian dissertation.

It is not surprising that the paper produced a strong reaction among Kekulé's friends. In the summer of 1865 Markovnikov came to Germany intending to work with Baeyer in Berlin. He wrote Butlerov of his reception there.

Baeyer asked me whether I had written a critique on the book of Kekulé, and when he received an affirmative answer, I met with a real battle. He reproached me for criticizing Kekulé as some sort of a bandit, and at the end he said, "I could wish that you would analyze the book of Kolbe in the same way, and you would see that there would be nothing left of it" (*25*).

Not surprisingly, Markovnikov decided to continue his journey. He spent some time in Heidelberg with Erlenmeyer, and he wrote Butlerov from there.

My paper on Kekulé has caused me much trouble. Everyone is convinced that it was written by you, as Erlenmeyer told me. It would be difficult to show that almost the same thing was written in my dissertation, because they could not read it. I told Erlenmeyer this, and he himself said that he had assured Baeyer to the contrary when he said that you wrote the paper. It is awkward even to speak of this to others (*26*).

Kolbe

In spite of the interest in Russians manifested by "Eremeich" as the Russians called Erlenmeyer among themselves, Markovnikov did not find him a very satisfactory adviser and soon moved to Leipzig where he studied analytical methods with Kolbe. He seems to have found this a more congenial place to work, for he wrote Butlerov from there:

I am fully satisfied with my visit here and regret that I tried to spend a semester in Berlin. Kolbe was especially attentive to all the workers, and, contrary to expectations, he did not behave like a commanding general at all. On the contrary, he was very glad to discuss and argue, and I have already succeeded several times in locking horns with him. He himself often approached and asked me how one should understand formulas which were not written by his method (*27*).

It is an interesting fact that both Markovnikov and Kolbe were strong opponents of Kekulé and his theories, though Markovnikov represented the most progressive thinking of his day while Kolbe adhered almost fanatically to the older radical theory and was even more bitter than Markovnikov in his attacks on the Kekulé school. In spite of their different outlooks, however, Markovnikov admired Kolbe greatly. He referred to him as "honored master" and tells us that in his discussions with the older man he was able to persuade him to stop using the equivalent weight of oxygen, 8, in his formulas and to accept the atomic weight of 16 (*16*).

In 1867 Markovnikov returned to Kazan, where he succeeded Butlerov in the chair of organic chemistry in 1868 when the latter was called to St. Petersburg. Markovnikov then prepared his doctoral thesis, "Materials on the Question of the Reciprocal Effect of Atoms in Chemical Compounds" (*20*), for which he received his degree in 1869. It was in this thesis that he presented much of his most important work.

He noted in the introduction:

I have frequently heard the question why the chlorine of acetyl chloride is much more easily replaced than the chlorine of ethyl chloride, although in both compounds the chlorine is attached to the same carbon atom. In view of such problems, it seemed to me that some fundamental details were lacking (21).

It was this problem which he set himself to answer in the most general terms possible. He stated his basic position in the words:

If any element is added to another, the character which it shows in its compounds depends not only on the individual properties which it shows in its separate state, but also on the properties of those elements with which it is combined. In turn, this latter element is subjected to the properties of the first, and this reciprocal effect is shown by the general behavior characteristic of the complex body (22).

He illustrated this by a number of examples in the behavior of chlorine, either alone or in such pairs of compounds as phosphorus trichloride and phosphorus pentachloride, phosgene and chloroform, and so on. His experimental studies, particularly on the manner of addition of halogen acids to hydrocarbons led him to numerous important generalizations, including the "Rule" for which he is best known to organic chemists.

Actually, it was his ability to generalize that led to his greatest contributions to chemistry. What he achieved was to move from the static study of chemical reactions by which Butlerov had shown it to be possible to establish a single formula for each compound to the deeper question of the mechanisms of chemical reactions and the factors which controlled them. The generalizations he made and the types of reactions he studied could not be explained satisfactorily until the electron theory of organic chemistry was developed in the 20th century. However, it is clear that it was by building on the foundations which he laid that this branch of chemistry has evolved. It was Markovnikov who initiated the third of the major developments mentioned at the beginning of this paper, the demonstration of the mutual interactions of the atoms in a molecule.

Conclusions

Thus we see that the decade from 1860 to 1870 was the turning point from the older to the modern view of organic chemistry. The first half of the decade was dominated by the study of aliphatic compounds, and without an understanding of their structure, Kekulé could not have produced his theory of the benzene ring, thus opening the field of aromatic chemistry. However, once the theory of chemical structure had been properly understood, it was inevitable that it would become the basis of organic chemistry. More and more chemists would begin to use it so that by the end of the decade it would become so universally accepted as to be taken for granted. This explains why the controversies of the decade gradually died out, and questions of priority ceased to be discussed. Yet it was the polemical literature of the sixties almost as much as the experimental work which engendered and hastened the final acceptance of the structural theory. Therefore, in following this literature we are following the main line of chemical development in this vital transitional period in the history of organic chemistry.

Acknowledgment

The author wishes to thank G. V. Bykov of the Institute of the History of Science and Technology of the Academy of Sciences of the U.S.S.R. in Moscow, who kindly supplied much of the material upon which this paper is based.

Literature Cited

(1) Butlerov, A. M., *Z. Chem.* **1861**, 549-560.
(2) Bykov, G. V., *Vopr. Istor Estestvoznaniya i Tekhn.* No. **4**, 179 (1957).
(3) Bykov, G. V., *Ibid.* No. **13**, 101 (1962).
(4) Bykov, G. V., Jacques, J., *Rev. Hist. Sci. Appl.* **13**, 123 (1960).
(5) Bykov, G. V., Sheptunova, Z. I., *Tr. Inst. Istorii Estestvoznaniya i Tekhn.*, *Akad. Nauk SSSR* **30**, 97 (1960).
(6) Conrad, M., *Ber.* **43**, 3645 (1910).
(7) Heintz, W. H., *Ann.* **132**, 1 (1864). The portion cited is on pp. 21-23.
(8) Kalinsky, B. A. and Bykov, eds., "Centenary of the Theory of Chemical Structure," pp. 44-53, Academy of Sciences Press, Moscow, 1961. Published simultaneously in Russian and English with French or German papers in the original languages. The references are to the English edition.
(9) *Ibid.*, p. 136.
(10) *Ibid.*, p. 140.
(11) *Ibid.*, pp. 102-110.
(12) Kekulé, A., "Lehrbuch der organischen Chemie, Erster Band," p. 152, Erlangen, 1861.
(13) *Ibid.*, pp. 156-157.
(14) Leicester, H. M., *J. Chem. Ed.* **17**, 203 (1940).
(15) Leicester, H. M., *Ibid.* **18**, 53 (1941).
(16) *Ibid.*, pp. 54-55.
(17) Leicester, H. M., *J. Chem. Ed.* **36**, 328 (1959).
(18) Markovnikov, V. V., *Z. Chem.*, n. f., **1**, 250 (1865).
(19) Plate, A. F., Bykov, G. V., eds., "V. V. Markovnikov, Selected Works," (in Russian), pp. 13-124, Academy of Sciences Press, Moscow, 1955.
(20) *Ibid.*, pp. 147-264.
(21) *Ibid.*, p. 150.
(22) *Ibid.*, p. 194.
(23) Plate, A. F., Bykov, G. V., Eventova, M. S., "Vladimir Vasil'evich Matkovnikov, Sketch of Life and Activities," p. 28, (in Russian), Academy of Sciences Press, Moscow, 1962.
(24) *Ibid.*, p. 43.
(25) *Ibid.*, p. 27.
(26) *Ibid.*, p. 28.
(27) *Ibid.*, pp. 29-30.

RECEIVED September 24, 1965.

3

Kekulé and the Dye Industry

DAVID H. WILCOX, JR.

Tennessee Eastman Co., Kingsport, Tenn.

August Kekulé worked on his theories for a long time before publishing them, and when he did publish, he was almost too late for priority. However, his theories were so well founded that they have withstood the real critical examinations of the ensuing years. The benzene theory was his greatest achievement. After its publication, Kekulé wrote nothing more in support and explanation until four and seven years later. One group of his former students, assistants, and friends kept the benzene theory alive and developed it through their own structural concepts and elaborations; the other group, intent on building a synthetic dye industry, made good use of Kekulé's work to convert a struggling industry enmeshed in alchemical empiricism into a fantastically large and important industry.

Organic chemistry found fulfillment and maturity in dyes, drugs, and myriad other valuable materials. Kekulé's benzene theory and subsequent elaboration of details replaced alchemical empiricism in dye making with the use of planned structures. By the untiring efforts of Kekulé's students and friends his theory was tested, modifications were suggested, and countless derivatives were prepared. This work kept the aromatic structure in the forefront and in use. When the impact of its utility was fully realized, the dye industry was not ungrateful. This turn enabled a slightly incredulous Kekulé to discern and enjoy during his lifetime the fruits of his theories, a feat not always accomplished by many forward-seeing thinkers.

Seventy-five years ago Kekulé said (*168*):

Here you came to celebrate the jubilee of the benzene theory. First of all let me say that this benzene theory was only a consequence, an obvious sequence, of my views; views which I held regarding the chemical value of elemental atoms, and of what we now regard as valence and structural theory. What else could I do with chemical affinities which remained disposable?

He could do nothing but link them together, and he had his benzene ring.

To Adolf Baeyer this theory was the keystone for building structural organic chemistry. Carl Glaser, speaking at the time for the German coal tar dye industry said (*168*):

The industry of artificial dyes derived from benzene and its derivatives, which come from coal tar, is an accomplishment of the last three decades. We can say with pride that the German industry undoubtedly occupies the first place in the world. In a victorious race our products have conquered the world market. We are no longer dependent on the dyes from madder and the various woods coming from abroad.

We are grateful for these surprising and remarkable successes of the home dye manufacturers, and we can ask ourselves: why did this industry especially develop in the German commonwealth? The answer is that the German universities were highly developed and staffed with distinguished teachers; among those I would like to name today: August Kekulé.

Courtesy Bildarchiv Foto Marburg

Figure 1.   Portrait of August Kekulé by Heinrich von Angeli

In grateful recognition of the cause of this prosperity of the German coal tar dye industry, we have commissioned a famous artist (Heinrich von Angeli) to paint the portrait of our celebrated scientist.

Near the end of his address, which followed, Kekulé turned again to Carl Glaser (168):

Your speech did not surprise me. A portrait cannot be made secretly. Even the viewing of the portrait later on will come as no surprise as it was made under my very eyes. My surprise goes back to an earlier time. During the fall vacation, spent atop of Rigi, I received your letter informing me of your intentions. Then two days later our friend Caro appeared to tell me personally. This was your desire to make it emphatic. This was the time that I was really surprised! For up to that time I was of the opinion that according to the views of the manufacturers, among whom I have many friends and former pupils, only the bee which carries and stores the honey has merit, but not the flower which produces the nectar.

That some of my researches and the benzene theory were of value to the technology of the coal tar dyes, I will not dispute, but I can assure you that I have never worked for industry, but only for science. I have always followed industry with a great deal of pleasure, but I have never accepted any compensation from it. It is for this reason that I am doubly delighted and doubly grateful that industry has seen fit to recognize my small merits.

Obviously, the dye industry did not feel that his contributions had been so small, and so I shall try to show some of the relationships between Kekulé and the dye industry.

Kekulé actually had many points of contact with the dye industry. Some he may have known while others he may have felt were quite indirect and farfetched. These areas of influence could include those listed below.

Direct Influence

Theory

Methane type
Tetrahedral carbon
Carbon to carbon linkage—chain structure
Carbon atoms linked together in a ring system—benzene structure
Other structures—e.g., pyridine type

Research which led to progress in the dye industry

Preparation of dye intermediates such as: phenol, isatin, dihydroxytartaric acid and anthraquinone
Structures of pyridine and triphenylmethane
Conversion of diazoaminobenzene into aminoazobenzene
Coupling of diazotized aniline with phenol to give an azo dye
Sporadic investigations on synthesis of indigo

Acted as consultant, law court expert and referee

Indirect Influence

Students and assistants

In dye industry

August Bernthsen
Heinrich Brunck
C. Bülow
Carl Glaser

Adolf Bruning
Carl Müller
A. Weinberg

In the universities, but who made dyes

Adolf Baeyer
Emil Jacobsen
Gustav Schultz

Otto Wallach
H. Wichelhaus

In the universities, but who elaborated and utilized the benzene theory

Richard Anschütz
F. Beilstein
J. Dewar
Francis R. Japp
Wilhelm Körner

August Ladenburg
August Mayer
Jacobus Henricus van't Hoff
J. F. Walker
Theodor Zincke

Friends in industry and universities

Henry E. Armstrong
St. Cannizzaro
Heinrich Caro
Carl F. Duisberg
E. Erlenmeyer
Carl Graebe
Peter Griess
A. W. Hofmann
Paul Julius

Ivan Levinstein
C. Liebermann
C. A. Martius
Victor Meyer
William Odling
Adolf Strecker
Otto N. Witt
Adolf Wurtz

The sum total of these influences was so enormous, it is not surprising that a benzene celebration was organized only 25 years after the publication of Kekulé's benzene papers (*168*).

There is not time to recall Kekulé's contributions to the theory of organic chemistry. They have been well covered in the past (*6, 116*) and in many papers down to the present (*183*).

Phenol

Some of Kekulé's research supplied the dye industry with valuable intermediates and working procedures. The first chemical was phenol, which had been obtained from coal tar by Runge (*163*) and had been named by Gerhardt (*82*). Still, it was a little more than a laboratory curiosity when Kekulé (*127*) and Adolf Wurtz (*188*) both came up with identical methods for its preparation by fusing benzene sulfonic acid with potassium hydroxide. Actually, the editor of the *Comptes*

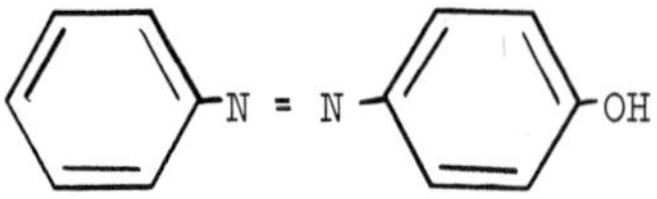

C. I. Solvent Yellow 7

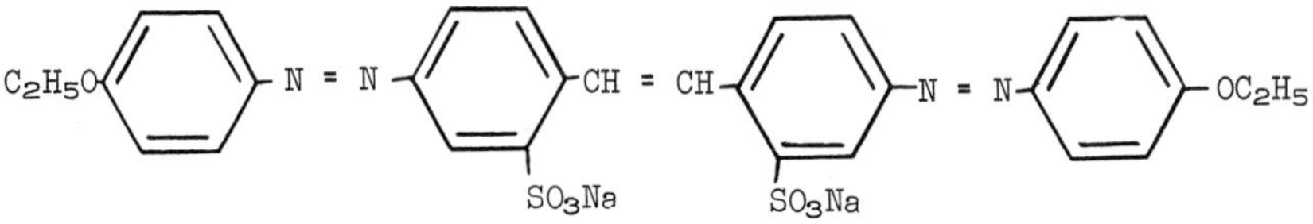

C. I. Direct Yellow 12
Chrysophenine

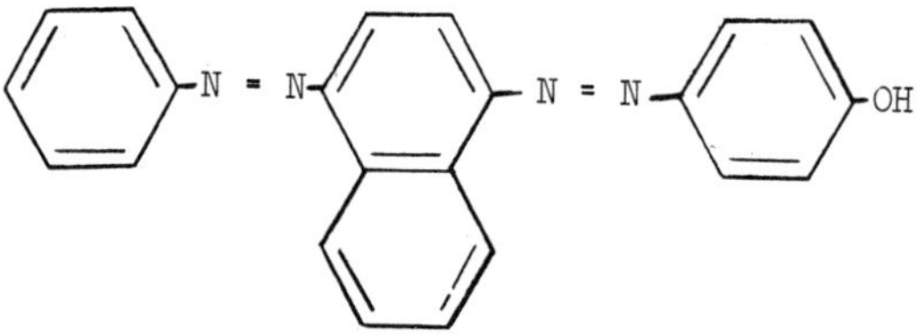

SRA Golden Orange I
C. I. Disperse Orange 13

rendus placed Wurtz's article just ahead of Kekulé's. However, I choose to place Kekulé first since he predicted the analogous formation of resorcinol—another valuable dye intermediate. These processes were practical and were used exclusively until recent years. The "Colour Index" (*173*) contains 60 dyes whose preparations use phenol as an intermediate. Examples range from C. I. Solvent Yellow 7 (4-hydroxy-azobenzene, the first azo dye prepared by Peter Griess), through C. I. Direct Yellow 12 (chrysophenine, made by ethylation of brilliant yellow) to C. I. Disperse Orange 13 (S.R.A. golden orange I) (*181*).

Resorcinol

Resorcinol was mentioned as a derivative of *m*-phenolsulfonic acid in Kekulé's 1867 phenol paper, but it was not until 1875 that Barth and

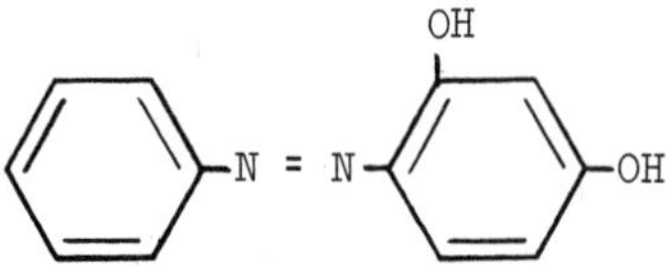

Sudan Orange G
C. I. Solvent Orange I

Senhofer (28) reported its preparation by this route. However, in the previous year they had reported its preparation directly from *m*-benzene-disulfonic acid (27). It has been commercially prepared this way for many years, even to this day. Resorcinol is still a component in 95 dyes

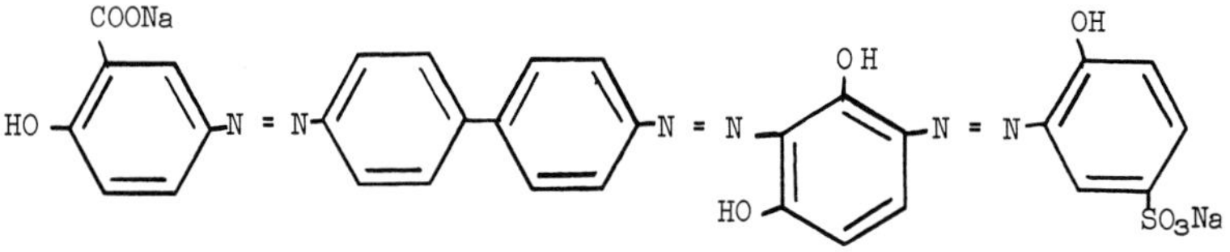

Sirius Fast Brown BRS
C. I. Direct Brown 95

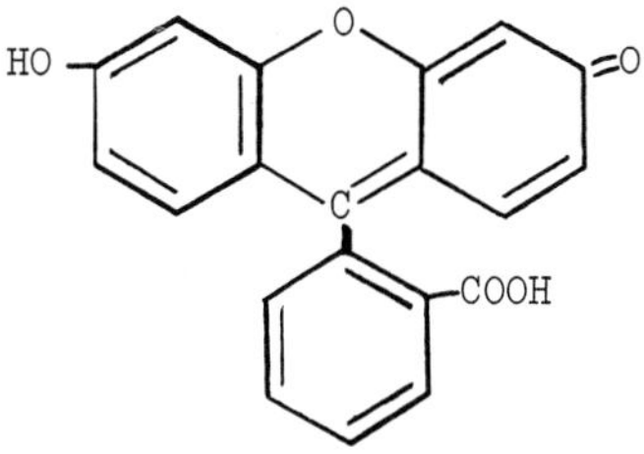

Fluorescein
C. I. Acid Yellow 73

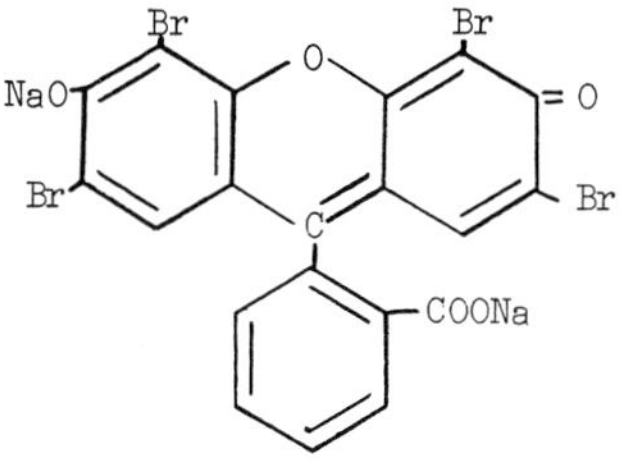

Eosine
C.I. Acid Red 87
C. I. Pigment Red 90

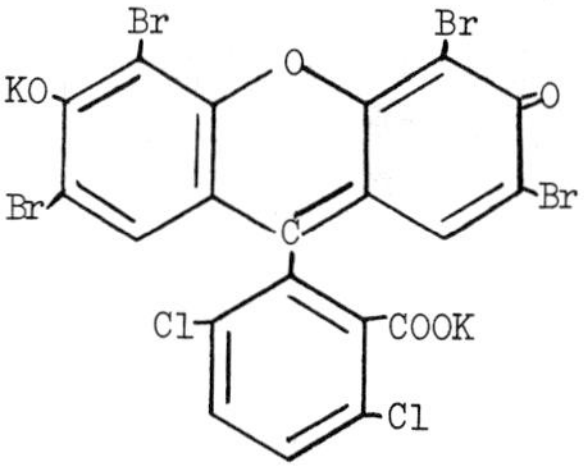

Phloxine
C. I. Acid Red 98

listed in the "Colour Index," ranging from Sudan orange, invented by Adolf Baeyer and C. Jaeger in 1875, through Sirius fast brown BRS, to fluorescein, discovered by Baeyer in 1871. Its sodium salt, Uranine, proved both spectacular and valuable as the sea marker dye during World War II. Fluorescein and its analogs and halogenated derivatives have spawned the fabulous eosines and phloxines.

Isatin

In 1869 Kekulé felt that he was making progress on the synthesis of indigo and is said to have asked Adolf Baeyer, his old pupil and assistant, to delay his work for a while (*124*). Through his interpretation of some of Carl Glaser's work, he had suggested the formulas for isatin and isatinic acid. However, Kekulé failed in his efforts to convert *o*-nitrophenylacetic acid into oxindole. After waiting for eight years,

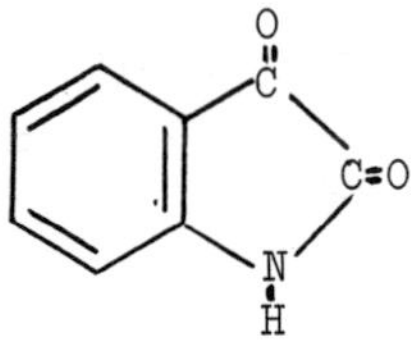

Isatin

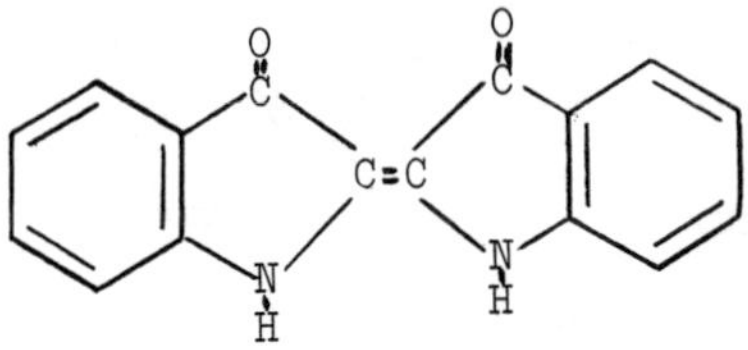

Indigo
C. I. Vat Blue 1

Baeyer resumed his research on the indigo synthesis. In 1870 Baeyer had chlorinated isatin with phosphorus pentachloride to obtain isatin chloride, which upon reduction was transformed into indigo. Baeyer was able to reduce *o*-nitrophenylacetic acid and convert it into oxindole, which could be oxidized to isatin. Isatin was first synthesized this way in 1878. It is still listed as the intermediate for preparing five indigoid dyes, the most important being synthetic Indigo, C. I. Vat Blue I.

Dihydroxytartaric Acid

Dihydroxytartaric acid had been obtained when protocatechuic acid, pyrocatechin, or guaiacol reacted with N_2O_3 in a solution of ether. Kekulé, in support of his benzene theory (*123*), showed that while

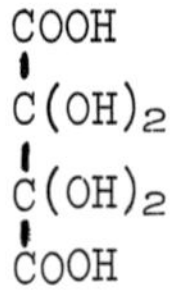

$$
\begin{array}{l}
COOH \\
| \\
C(OH)_2 \\
| \\
C(OH)_2 \\
| \\
COOH
\end{array}
$$

Dihydroxytartaric Acid

dihydroxytartaric acid could be formed from these benzene derivatives, this did not prove that in benzene one carbon atom was linked with three other carbon atoms. He went on to indicate that the substance which was supposed to be "carboxytartronic acid" could be made from nitrotartaric acid by the action of nitrous acid in alcohol solution and that this substance could be converted on reduction into racemic and meso-tartaric acid. Kekulé named "carboxytartronic acid" dihydroxytartaric acid.

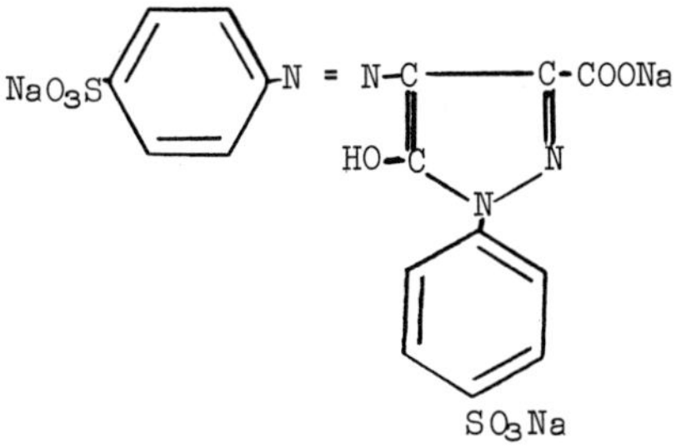

Tartrazine
C. I. Acid Yellow 23
C. I. Food Yellow 4

J. H. Ziegler in 1884 added two moles of *p*-hydrazinobenzene sulfonic acid (*72*) from diazotized sulfanilic acid (*165*) to dihydroxytartaric acid to form tartrazine (*11, 117, 189, 190*).

It was Richard Anschütz, pupil of Kekulé and his heir at Bonn, who pointed out that when the phenylhydrazine condensed with the dihydroxytartaric acid, an unstable osazone was first formed, which then dehydrated to give the pyrazolone (*3*).

Pyridine

Pyridine is the basic ring structure of many plant alkaloids. Wilhelm Körner, a pupil of Kekulé, proposed the following structural formula (*139*).

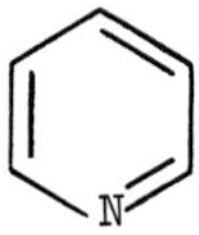

Pyridine

Riedel proposed that the nitrogen was bonded to three carbons, and Bamberger and Pechmann suggested a centric formula. Kekulé (*157, 183*), seeing the same kind of disagreement that had challenged his own structure for benzene, set up a series of experiments to settle the matter. He proved (*157*) the presence of the imido group by converting glutaconimide into methyl glutaconimide and preparing nitrosoglutaconimide. He converted glutaconimide (*2*), and pyridine, with phosphorus pentachloride into the same pentachloropyridine. These

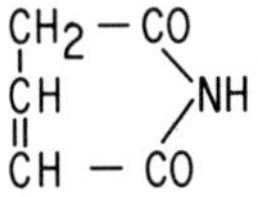

$$CH_2 - CO$$
$$CH \qquad NH$$
$$CH - CO$$

GLUTACONIMIDE

experimental facts supported the Körner structural formula. While pyridine has not until recently been a component of dyes—e.g., new basic dyes for acrylic fibers—it has been helpful both as a reaction medium and as a catalyst.

Triphenylmethane

Triphenylmethane, the nucleus of the great group of colors ranging from the early rosaniline dyes to malachite greens, aurines, and phthaleins, was first prepared by Kekulé in 1872 by the action of benzal chloride on mercury diphenyl (*126*). Emil and Otto Fischer, colleagues of Kekulé's pupil, Baeyer, tried diazotizing the leuco form of *p*-rosaniline (prepared from aniline and *p*-toluidine) and thereby obtained the hydrocarbon, $C_{19}H_{16}$ (*75, 158*), identical with Kekulé's triphenylmethane. They then nitrated the hydrocarbon to the trinitro derivative, which

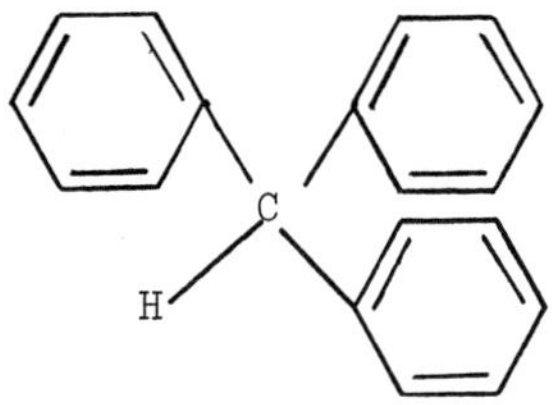

Triphenylmethane

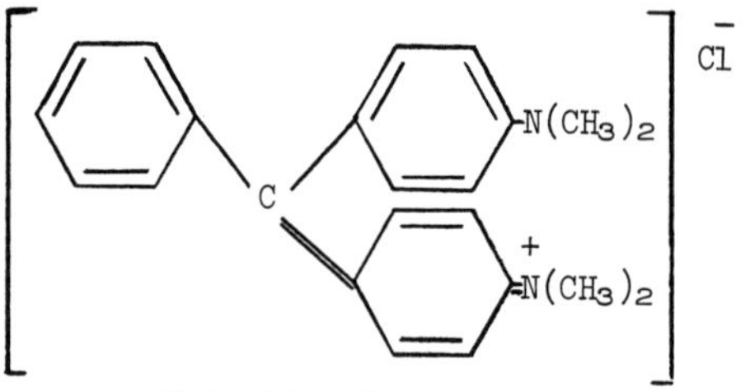

Malachite Green
C. I. 42000
C. I. Basic Green 4
C. I. Pigment Green 4
C. I. Solvent Green 4

upon reduction gave the triaminotriphenylmethane, also known as paraleucaniline. When the hydrochloride of the amino compound, was heated to 150°–160°C., it was transformed into *p*-rosaniline (*73*). Experiments like these clarified the constitution of this important class of dyes (*156*).

Diazoamino–Aminoazobenzene Rearrangement

In the sections on diazo and azo compounds the nomenclature and structures will be those used in the original papers.

In 1858 Peter Griess (*180*) prepared the first diazo compound, diazodinitrophenol (*100*), by bubbling nitrous acid gas through a cold alcoholic solution of picramic acid. This new compound, with interesting properties, was followed by many others (*101*). Later that year Hofmann reported that Griess had, by the action of nitrous acid on phenylamine (aniline), obtained a new "fusible body," (*106*) $C_{24}H_{11}N_3$, which is insoluble in water and easily soluble in alcohol. This compound, which possesses weakly basic characteristics, is formed according

to the following equation (note that molecular weights C = 6 and
O = 8 were still in use):

$$C_{24}H_{14}N_2 + NO_3 = 3\ HO + C_{24}H_{11}N_3$$

| 2 equivalents of | new |
| phenylamine | compound |

This was later named diazo-amidobenzol (*102, 103*).

In 1864 Griess reported the first azo dye: phenylazophenol and the
first disazo dye: bisphenylazophenol (*107*). In this communication
Griess wrote:

Chemists are not agreed upon the rational constitution of amido-
compounds. They are frequently referred to the ammonia-type and
almost as frequently to the same type to which the nitro-compounds,
from which they are derived, belong. In the latter case the group NH_2
is considered as replacing one atom, or NH_3 as taking the place of two
atoms of hydrogen. Aniline can thus be written in three different ways
and expressed by the three formulas,

$$\left.\begin{array}{l} \Theta_6H_5 \\ H_2 \end{array}\right\}N \qquad\qquad \Theta_6H_5(H_2N) \qquad\qquad \Theta_6H_4(H_3N)$$

| Phenylamine | Amidobenzol | Ammoniabenzol |

[Griess and Kekulé used Θ and Θ to represent atomic weights 12
and 16, respectively.] The two latter formulas appear to be capable of
explaining in the most natural manner the formation of bodies in which
nitrogen is substituted for hydrogen.

Speaking of these diazo derivatives of amines he said:

They are remarkable for the great variety of compounds which
they produce, such as is not met with in any other portion of the field
of organic chemistry. . . . Altogether they may be looked upon as one
of the most interesting groups of organic compounds. I have avoided,
as much as possible, discussing their rational composition and have
abstained from theoretical speculation. I have, however, come to the
conclusion that the two atoms (or the molecule) of nitrogen, N_2, they
contain must be considered as equivalent to two atoms of hydrogen,
and it is in accordance with this view that the names of the new compounds have been framed.

In the meantime, a new dye appeared, called "aniline yellow," which
came from the dye plant of Simpson, Maule, and Nicholson (*80, 145*).
Perkin said that Nicholson had prepared it by an unpublished process
(*151*) and that Dale and Caro obtained a patent to make it in 1863 (*58*).
Obviously, he says, "the dye's constitution was unknown, and its preparation was empirical. It was not a very successful dye because of its
volatility." However, it was listed as No. 22 in Schultz and Julius' first
dye table (*169, 170*), but it was recorded in Green's "Organic Colouring
Matters (1904)" "as a dyestuff no longer in commerce" (*99*). Shortly
after it was on the market Peter Griess examined and identified it as

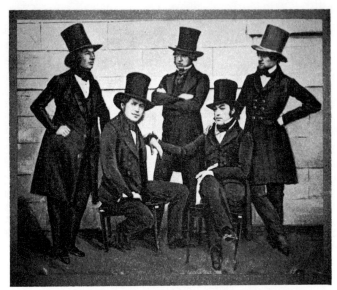

Courtesy of Imperial College of Science and Technology,
London, from Henry Edward Armstrong by J. Vargas Eyre

*Figure 2. Hofmann and others at Geissen University about
1842. Left to right: Karl Renigius Fresenius, Heinrich Will,
John Lloyd Bullock, John Gardner, August Wilhelm von
Hofmann*

the product obtained when his "diazo-amidobenzol" rearranged on
standing in the presence of acid. This rearrangement set off a lively
theoretical and experimental controversy which has engaged many chemists for a long time, and it is not dead yet.

In 1865 Kekulé, while working on his textbook, had arrived at the
point devoted to the aromatic diazo and azo compounds. This section,
Anschütz states, was covered "in paragraphs 1729–1769 which conclude
the second volume, in a most exemplary manner its theoretical foundation" (132, 133). At the same time he prepared two excellent treatises.
Kekulé introduces his work "On the Constitution of Diazo Compounds"
(1, 131):

For a long time new substances have attracted chemists, but not
to such a high extent, and quite justly so, as the diazo compounds discovered by Griess. As far as the constitution of these peculiar substances
is concerned, many suppositions have been made, but it appears to me
that none of these fits the facts or explains them in a satisfactory manner.

Kekulé then proceeded to explain his concept of the mechanism as
it fitted in with his new benzene theory (47, 70, 71, 154):

Griess himself, as he mentions in his papers, had avoided all theoretical considerations. He pointed out that one can compare diazo com-

pounds with amino compounds, from which they are prepared, or with the substances from which these same amino derivatives are obtained. Then 3 atoms of hydrogen of the amino derivative are replaced by one atom of nitrogen, or 2 hydrogen atoms of the normal substance are replaced by the equivalent group, N_2, for instance:

$\mathrm{C_6H_4 \cdot H_3N}$	$\mathrm{C_6H_4N_2}$
Aniline	Diazobenzol
$\mathrm{C_6H_6}$	$\mathrm{C_6H_4N_2}$
Benzol	Diazobenzol

The constitution of this two-valent group (azo) can be easily derived from the fundamental principles of the atomic theory, as was pointed out by Erlenmeyer and Butlerov: —N≡N—.

Kekulé further pointed out that it is difficult to assume, in accordance with his benzene theory, that the two-valent group, N_2, replaced two hydrogen atoms of benzene (*131*).

The hydrogen atoms occupy, in the benzene ring, non-adjacent positions, and a substitution of two non-adjacent hydrogen atoms by a two-valent atom is most improbable. I presume, therefore, that the two-valent, N_2, group can only be linked with the carbon of benzol in one position and that in all simple diazo benzol derivatives five hydrogen atoms remain.

The most simple compounds of diazobenzene, for instance, may be expressed by the following formulas:

Diazobenzolbromid	$(\mathrm{C_6H_5})$—N≡N—Br
Diazobenzolnitrat	$(\mathrm{C_6H_5})$—N≡N—NO₃
Diazobenzolsulfat	$(\mathrm{C_6H_5})$—N≡N—SO₄H
Diazobenzolkali	$(\mathrm{C_6H_5})$—N≡N—OK
Diazobenzolsilberoxyd	$(\mathrm{C_6H_5})$—N≡N—OAg
Diazobenzolamidobenzol	$(\mathrm{C_6H_5})$—N≡N—NH$(\mathrm{C_6H_5})$.

Griess assumed the existence of a free diazobenzene. However, it is most likely:

Diazobenzolhydrat $(\mathrm{C_6H_5})$—N≡N—OH.

From his interest in the theory of diazo compounds Kekulé became interested in phenolsulfonic acids (*130*). By the action of concentrated sulfuric acid on diazobenzene sulfate, Griess had obtained an acid which he called "disulfophenylensäure" (*104*). Kekulé explained (*131*) that this compound was nothing but phenoldisulfonic acid and was formed according to a two-step reaction:

	Before Decomposition	*After Decomposition*
1st step	$\dfrac{\mathrm{C_6H_5 \cdot N_2 \cdot HSO_4}}{\mathrm{OH \cdot H \cdot SO_3}}$	$\mathrm{C_6H_5} \Big\vert\ \mathrm{N_2}\ \Big\vert\ \mathrm{HSO_4}\ \Big\vert\ \mathrm{SO_3}$ $\mathrm{OH}\ \ \ \ \ \ \ \ \ \ \ \ \ \ \ \ \mathrm{H}$
2nd step	$\mathrm{C_6H_3 \cdot OH + H_2SO_4 \cdot SO_3}$	$\mathrm{C_6H_3}\begin{cases}\mathrm{OH}\\\mathrm{SO_3H}\\\mathrm{SO_3H}\end{cases} + \mathrm{H_2O}$

To test the correctness of his views, Kekulé together with Leverkus, prepared phenoldisulfonic acid by the action of fuming sulfuric acid upon phenol (130). This product was found to be identical with Griess' "disulfophenylensäure."

Cyclic Diazoamino Compounds

By using o-phenylenediamine, which contains amino groups in adjacent positions, the formation of specific closed-ring diazoamino compounds is possible. Such cyclic diazoamino compounds were obtained by Hofmann, by treating nitrophenylenediamine with nitrous acid (113). Kekulé reported (131):

The analogy of such diazoamino derivatives with the diazoamino compounds from monoamine derivatives can be distinctly seen from the following formulas:

$$
\mathrm{C_6H_5N} \atop \mathrm{C_6H_5NH} \Big\}N \qquad \mathrm{C_6H_3(NO_2)}\Big\{ {N \atop NH} \Big\}N \qquad \mathrm{C_6H_4}\Big\{ {N \atop NH} \Big\}N
$$

| Diazoaminobenzene | Nitrodiazodi-
aminobenzene | Diazodiamino-
benzene (140) |

in which Ladenburg described the diazoaminobenzene obtained from o-phenylenediamine with nitrous acid which he called "aminoazophenylene."

We now call this 1,2,3-benzotriazole.

Constitution of Diazo Group

These explanations sufficed for a while, but before long some questions developed concerning the diazo group, —N=N—. Kekulé, a firm believer in a trivalent nitrogen, felt that any other formula for the diazo group was unthinkable. Nonetheless, those favoring a changing valence for nitrogen as either a three- or five-valent element, continued to express their views. Blomstrand saw another possibility and gave to the diazo group the formula, —N—, in which one nitrogen atom was

$$\overset{\displaystyle |||}{N}$$

pentavalent. According to his view, it was similar to ammonium compounds, which Kekulé considered to be molecular addition compounds. Independently of Blomstrand, Adolf Strecker (175, 176) in 1871 investigated the reduction of diazo compounds with sodium acid sulfite and concluded that the diazo group has the formula —N— . In 1874 Erlen-

$$\overset{\displaystyle |||}{N}$$

meyer (69) agreed with Blomstrand. Blomstrand (40) continued to claim priority for his concept of diazo compounds as types similar to ammonium compounds.

Phenylhydrazine

Adolf Strecker, as mentioned above, had prepared a derivative of phenyl hydrazine: potassium phenyl hydrazine *p*-sulfonate (*175*). This was the first aromatic hydrazine, and he called it, "diazid der sulfanilsaure."

Anschütz wrote:

Emil Fischer [72, 74] in his excellent piece of research, isolated phenyl hydrazine. He supported the viewpoint of Kekulé for the formula of the diazo compound as, —N=N—. This he based upon the transformation of the diazo compound into phenyl hydrazine, to which he definitely gave the formula: $C_6H_5NH\cdot NH_2$. The formation of mixed azo compounds can be easily explained with this formula without assumption of a rearrangement.

Azobenzene

Carl Glaser and Kekulé reviewed the manner in which Mitscherlich converted nitrobenzene with alcoholic potassium hydroxide into azobenzene in 1834. They felt that it was a reduction process, in which nascent hydrogen combined with oxygen of nitrobenzene to form water. The remaining free nitrogen affinities left over from two molecules then joined to form azobenzene. They reasoned that azobenzene could be obtained in the reverse manner from aniline nascent oxygen's combining with the hydrogens of the amino compounds, which would then allow the free nitrogen affinities of the two molecules to unite. Glaser then prepared azobenzene by oxidizing aniline and obtained small amounts of azoxybenzene and traces of hydrazobenzene. However, potassium permanganate appeared to be specific for this oxidation. It will be recalled that Perkin made mauve by oxidizing impure aniline with potassium dichromate—obviously a difficult oxidation. Glaser gave the following equation for the oxidation:

$$2C_6H_5\cdot NH_2 + O_2 = C_{12}H_{10}N_2 + 2H_2O$$

He also reduced his azobenzene with alcoholic ammonium sulfide to hydrazobenzene, which upon heating in dilute sulfuric acid, rearranged into benzidine sulfate (*86, 89*).

Kekulé then published his paper on "Relation between Diazo and Azo Compounds and the Transformation of Diazoaminobenzene into Aminoazobenzene" (*129*). In the introduction, Kekulé pointed to Glaser's paper, in which their views on the constitution of azobenzene were reported, and made the following statement:

If we compare the formula of azobenzene with that of diazobenzene bromide or its corresponding compounds:

Azobenzene (C_6H_5)—N=N—(C_6H_5)
Diazobenzene bromide (C_6H_5)—N=N—Br

It can easily be seen that both formulas have one part in common:

$$(\text{C}_6\text{H}_5)\text{—N}\text{=}\text{N}\text{—}$$

This is our phenylazo radical.

Aminoazobenzene

Kekulé wrote (129):

Diazoaminobenzene is isomeric with the base "amidodiphenylimid" [145], and I have shown previously [128] that it is formed by the action of bromine upon aniline.

Kekulé preferred the name aminoazobenzene to "amidodiphenylimid" since one would have to assume that two benzene residues are held together by carbon valences. Kekulé reminds us that his formula for aminoazobenzene coincides with those of Fittig (76) and Erlenmeyer (71).

Griess and Martius (145) had found that in the action of nitrous acid upon an alcoholic solution of aniline the temperature determined whether diazoaminobenzene or aminoazobenzene is formed. These observations induced Kekulé to study both compounds (129) and he observed:

The above mentioned views appear to make it probable that under suitable conditions diazoaminobenzene goes over into aminoazobenzene. I have found that this transformation occurs easily and completely if one allows diazoaminobenzene to remain in alcoholic solution in the presence of aniline hydrochloride for a little while. The reaction may be expressed by the following equation:

$$\text{C}_{12}\text{H}_{11}\text{N}_3 + \text{C}_6\text{H}_7\text{N}\cdot\text{HCl} = \text{C}_{12}\text{H}_{11}\text{N}_3 + \text{C}_6\text{H}_7\text{N}\cdot\text{HCl}$$

Diazoaminobenzene   Aminoazobenzene

It is by the transformation of diazoaminobenzene into aminoazobenzene that an equal amount of aniline hydrochloride is formed. It is obvious that a relatively small amount of aniline hydrochloride is sufficient to transform a large quantity of diazoaminobenzene into aminoazobenzene, and this has been confirmed by experiment. The aniline hydrochloride acts as a "fermenter."

Azo Compounds

Kekulé with Coloman Hidegh in 1870 reported some additional work on the constitution of azo compounds (125). They reviewed the conversion of diazoaminobenzene into aminoazobenzene and assumed that the same type of mechanism would hold true for hydroxy compounds. They reasoned that diazobenzene would act on phenol to give diazohydroxybenzene, an analogue of diazoaminobenzene. Through a molecular rearrangement it would then convert to isomeric hydroxyazobenzene.

Diazohydroxybenzene $\text{C}_6\text{H}_5\text{—N}\text{=}\text{N—O}\cdot\text{C}_6\text{H}_5$

Hydroxyazobenzene $\text{C}_6\text{H}_5\text{—N}\text{=}\text{N—C}_6\text{H}_4\cdot\text{OH}$

Kekulé added a nitric acid solution of diazobenzene nitrate gradually to an aqueous solution of potassium phenolate. The brown resinous body which formed soon hardened and proved to be identical with Griess's "phenoldiazobenzol" (*102*).

Kekulé as Consultant

Kekulé, like so many academic chemists today, acted as a consultant and legal expert and referee in patent trials. Richard Anschütz (*1*) and Arthur Weinberg (*179*) are the sources for the following examples.

Methylene Blue Case. Kekulé was involved in the methylene blue, C.I. 52015, case. Heinrich Caro had discovered the dye in 1876, and Badische patented it in 1877 as British patent 3751 (*50*) and German patent 1886 (*14*). He had prepared it by oxidizing dimethyl-*p*-phenylenediamine in the presence of hydrogen sulfide. Hoechst was granted a patent (*112*) eight years later for a blue dye which formed when a mixture of dimethyl-*p*-phenylenediamine and dimethylaniline was oxidized in the presence of a thiosulfate. Its constitution, however, was uncertain. August Bernthsen (*36*) in the meantime had carried out his splendid research on the structure of thionine (Lauth's Violet). As a result, Badische applied for a patent to make methylene blue from dimethyl-*p*-phenylenediamine thiosulfate. Two firms, G. C. Zimmer and Hoechst, objected by declaring that the disclosure of the thiosulfate method in the Hoechst patent made invalid any further inventions and that reducing a well known process into steps could hardly be considered a patentable subject. The Zimmer firm then produced consultant, Johannes Wislicenus, who defended this viewpoint. Then Badische presented as its consultant, August Kekulé, who had checked Bernthsen's experimental work. He told the court that the patent law had been created to promote and protect progress. He, too, felt that a distinct synthesis of a valuable product constituted progress and, therefore, constituted a patentable invention. When the patent was granted to Badische, the firms of Zimmer and Hoechst appealed the case, using new testimonies by Johannes Wislicenus and Emil Fischer. The patent office, nonetheless, continued to agree in all points with the views of Kekulé.

Antipyrine Case. Ludwig Knorr synthesized antipyrine in 1883 (*111, 134, 135, 136*). Phenylhydrazine was condensed with ethyl acetoacetate to form phenylmethylpyrazolone, which on subsequent methylation yielded the product; the process was patented by Hoechst (*110*). In 1890 the firm of Riedel applied for a patent for an antipyrine process, which consisted of heating together equivalent parts of phenylhydrazine, acetoacetic ester, sodium methyl sulfate, and sodium iodide along with a small amount of hydriodic acid. Naturally, Knorr and Hoechst objected, submitting two testimonies—one by Kekulé and the other by Emil Fischer. The latter covered the entire historical background leading up to and including the preparation of antipyrine by Knorr. Kekulé testified that Riedel's procedure did not represent a technical advance but was rather a step backwards, when compared with the step-by-step synthesis of Knorr. Nonetheless, the patent was issued, based on the justification that it was new and surprising that by Riedel's method the condensation and subsequent methylation had occurred. In the interference proceedings Kekulé was again the consultant, and this time the German patent office lost the case. Kekulé's closing remark was: "As far as he, Kekulé, was concerned the only thing new and surprising for him was that the patent office could believe such nonsense." The patent was not issued.

Chrysamine G Case. Chrysamine, C.I. 22250, prepared by coupling tetrazotized benzidine with two moles of salicylic acid, was patented by Bayer (30, 31, 77, 78). The Oehler factory at Offenbach came out in 1888 with a homologous dye, Kresotine yellow G, C.I. 22410 made from o-crestotinic acid instead of salicylic acid (147, 148, 162). Bayer brought suit in the court at Darmstadt, which recognized that there had been a violation of the Bayer patent, and accepted two consultants: Heinrich Caro as technical advisor and A. W. Hofmann as scientific representative. However, Hofmann died before finishing his scientific testimony, and Kekulé was appointed to take his place. Caro's testimony ran to more than 100 pages, proving both tedious and difficult for the lawyers to understand. Kekulé's testimony, on the other hand, was clear and concise, amounting to three pages, in which he indicated that the Oehler dye was undoubtedly a violation or infringement of the Bayer patent. In rebuttal Oehler called in Adolf Baeyer as consultant. Baeyer argued that nobody could be stopped from using cresotinic acid since it was not mentioned in the patent. However, Baeyer devalued his own opinion by his closing remarks: "In giving this testimony, I would like to point out that I am, when it comes to patent laws, not competent to render judgement." Hence Baeyer's testimony was unacceptable to the court, and a decision was rendered according to Kekulé's view. It should be noted, however, that Oehler eventually obtained patents in Germany, England, and the United States and that Bayer obtained a U. S. patent slightly ahead of Oehler (173).

Kekulé's Pupils, Assistants, and Friends

Space will permit only a brief note about a few of Kekulé's pupils, assistants, and friends—the area of his greatest influence. These chemists carried on Kekulé's work when affliction slowed him down.

Adolf Baeyer. Johann Friedrich Adolf von Baeyer was born October 31, 1835 in Berlin, the son of Johann Jacob Baeyer, a major general on the Prussian general staff. At an early age Baeyer started chemical experimentation, so that as an act of self-preservation his father gave him, on his ninth birthday, a copy of Stockhardt's "School of Chemistry."

Courtesy *Chemische Berichte*

Figure 3. Adolf von Baeyer

Baeyer began his chemical training under Robert Wilhelm Bunsen in Heidelberg. Here he completed work on two papers: one on idiochemical induction (15) and the other on methyl chloride (16). After two semesters with Bunsen, he went to work for two years in Kekulé's private laboratory in Heidelberg. A description of this laboratory by Bernthsen is of interest (35):

It was necessary for Kekulé to furnish his own small laboratory and auditorium, and the State "most generously" provided the benches for the students, which contained initials deeply cut into the wood. The house at 4 Haupstrasse had three windows on the front. It was owned by flour dealer, Goos. On the first floor were the living and bedrooms and on the second floor the lecture room and laboratory.

Baeyer said that the laboratory was exceedingly primitive. It consisted of a room having one window and two laboratory tables. There was no hood or any provision for a fume chamber. The kitchen was used as the room for evil-smelling gases, and its stove had very little draft (35). Here Baeyer continued his research on cacodyl compounds and discovered arsenic methyl chloride. Inhaling this compound caused him to faint and nearly cost him his life (153). On completing the work, Baeyer returned to Berlin and presented his thesis for a Ph.D. in Latin, as was the custom (17). Baeyer followed Kekulé to Ghent in the Winter 1858–59, and while on the way from Heidelberg to Ghent he met Adolph Schlieper. The latter had worked on uric acid under Justus Liebig and gave Baeyer a box of preparations, which Baeyer promptly began to investigate at Ghent.

In the Spring of 1860 Baeyer returned to Berlin as a teacher of organic chemistry in the Gewerbe Institut, on Klosterstrasse (later the Technische Hochschule, Charlottenburg). For 12 years he stayed in this poorly paid and modest position and had in his laboratory such men as Graebe, Liebermann, and Victor Meyer as recompense.

Baeyer moved to Strassburg in 1872, and there Emil and Otto Fischer were his pupils. He left early in 1875 to succeed Liebig at Munich. Apparently, Liebig had no desire "to continue the famous tradition of the Giessen School," (153) for Baeyer found that there was no teaching laboratory in the building that Liebig had built. The large new

Courtesy Liebig Museum, Giessen

Figure 4.   Justus Liebig's chemical laboratory at Giessen

laboratories planned by Baeyer were completed in 1877. Baeyer's own private laboratory, according to Perkin (*153*):

was equipped with the simplest possible apparatus, the most striking feature being large racks, such as are commonly in use at the present time, filled with test-tubes. He insisted always on absolute cleanliness, and it used to be said that his test-tubes were first soaked in dilute caustic soda for an hour, then washed with water, then with alcohol and finally with distilled water and dried—Baeyer was always conducting experiments—he gave only elementary lectures and was not very much interested in theory—there was a feeling in the lab that no one was of any account who did no research—an atmosphere which produced the greatest chemists of the day and weeded out those who were of no account.

Baeyer occupied the chair in Munich until his death at 82 on August 20, 1917. He received the Nobel Prize in 1905. From the bulk of his research, it is possible to select only one example—the analysis and synthesis of indigo.

INDOLE. According to Perkin (*153*), Baeyer said that his original impulse to work on indigo could be traced to an incident which occurred in his youth. On his 13th birthday he was given a two-thaler piece, with which he bought a lump of indigo. Young Baeyer became immediately fascinated with the properties of this material, and this fascination remained until he had solved the chemistry of the coloring matter. His experimental work on indigo began in 1865 (*25, 26, 79*).

Laurent and Erdmann had oxidized indigo and obtained isatin. Baeyer then showed that isatin on reduction gave dioxindole, which then could be converted into oxindole. When this compound was distilled with zinc dust, it changed to indole.

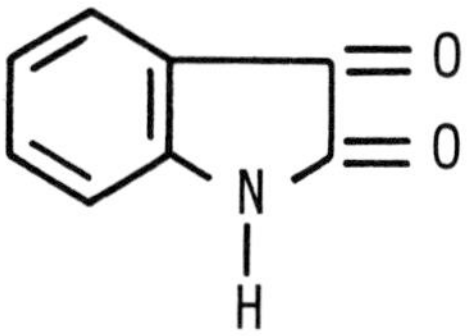

Isatin 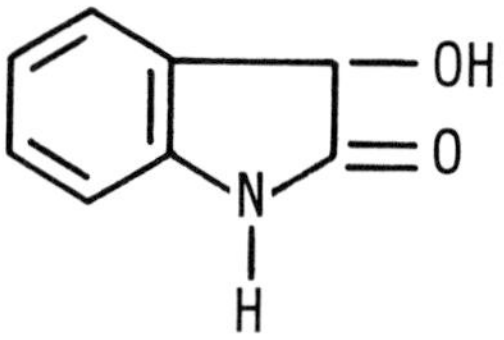   Dioxindole

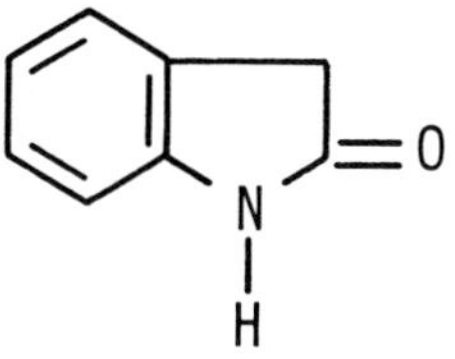

Oxindole 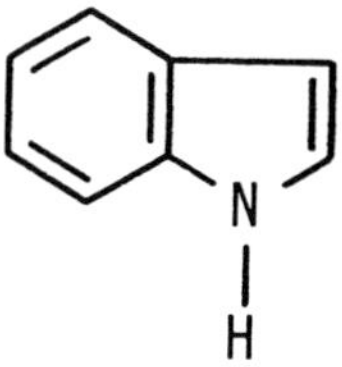   Indole

Baeyer and Emmerling (*18*) fused *o*-nitrocinnamic acid with potassium hydroxide and iron turnings to prepare indole. In 1870 Baeyer obtained indole directly from indigo by reductive distillation with zinc dust. Kekulé (*124*) at the time was interested and proposed the correct formulas for isatic acid and isatin but the incorrect one for indole. The correct formula for indole was given by Baeyer and Emmerling (*18*). These two then obtained indigo by heating isatin with phosphorus trichloride and phosphorus, and C. Engler and Emmerling obtained it from nitroacetophenone (*68*). At this point Baeyer discontinued his research in deference to the wishes of Kekulé, who felt that he could prepare, from *o*-nitrophenylacetic acid, *o*-nitrophenylacetylene which might cyclize to isatin. However, he was unsuccessful, and after about eight years, Baeyer resumed his indigo work. By using Kekulé's general idea, Baeyer was able to convert *o*-nitrocinnamic acid into *o*-nitrophenylpropiolic acid, which on heating, lost carbon dioxide to form isatin (*21, 22, 24*). However, on heating in a solution of alkali and reducing sugar, the blue needles of indigo were formed (*23*). In 1883 Baeyer published the first correct formula for indigo (*25*):

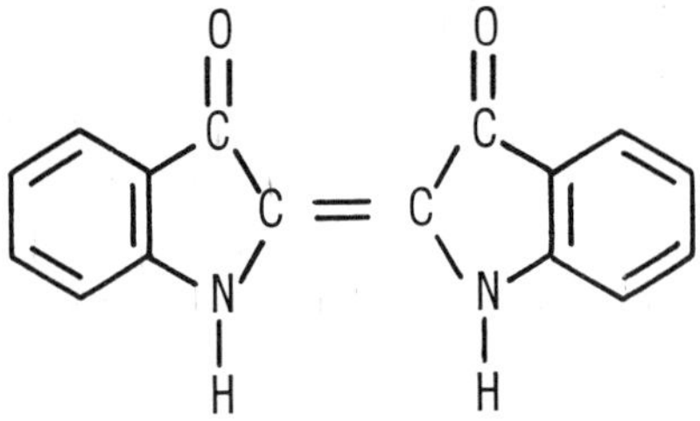

Indigo

Several routes to indigo quickly appeared (*19, 20, 177*), but they all were based on benzene or toluene, which were then scarce.

Like Kekulé, Baeyer produced many outstanding industrial dye chemists including: A. Spiegel, A. v. Weinberg, C. Duisberg, G. v. Bruning, B. Homolka, F. Stolz, V. Villiger, and B. Graf Schwerin (*46, 149, 153*). Hans Aickelin, head of General Aniline prior to World War II, told me that he was Baeyer's last graduate student.

Carl Graebe. Carl Graebe was born February 24, 1841 in Frankfurt am Main and died there on January 19, 1927. The years between were packed with an amazing amount of synthetic organic chemistry. He was Bunsen's pupil at Heidelberg, graduating in 1862. Graebe remained as Bunsen's lecture assistant for three sessions and then went to Berlin. Here he served as Baeyer's assistant from 1865 to 1869. Then he was instructor and professor at Leipzig for a year at Königsberg during 1870, moved on to Zurich for eight years, and from 1878 to 1906 he was at Geneva. He returned to Frankfurt in 1906 and lived there in retirement for the next 21 years. Graebe, during his retirement, wrote "Geschichte der Organischen Chemie." However, only one volume was published at Berlin in 1920. He also wrote memoirs about his friends, Marcelin Berthelot (*98*) and Adolf Baeyer (*90*).

TURKEY RED. While he was with Baeyer, Graebe successfully carried out the research which culminated in synthesizing alizarin, the main constituent of the coloring matter obtained from the madder plant. This was the first naturally occurring dye to be correctly analyzed and syn-

Courtesy M. Frosch, Badische

Figure 5.   Carl Graebe

thesized. The Badische chemists—Carl Glaser, Heinrich Brunck, and Heinrich Caro—combined their talents to develop a practical process which quickly ended the cultivation of the madder plant. A brief review of the background and advances in this area are of interest.

Turkey red dyeing with madder was discovered in India, but eventually the process became known to the Turks who brought it to the Near East—to Greece, Cyprus, and Smyrna. The town of Adrianopolis became famous for the fine red dyeings it produced. In 1742 Greek dyers brought the process to France, and from there it quickly spread to Alsace, Switzerland, and Germany. The early French and Italian dyers called it "Adrianopole Red."

Turkey red dyeing was established in Glasgow by Macintosh and Papillon about 1780. The process was complicated and tedious. Sansone (164), in 1885, stated that alizarin began replacing madder in the Turkey red dye works in 1871–72, and by 1873 the Swiss dyers were using only synthetic alizarin. Alizarin and purpurin were obtained from madder color in 1826 by Colin and Robiquet (159); the latter analyzed alizarin and came up with the formula $C_{37}H_{48}O_{11}$ (160). Edward Schunck, Liebig's pupil, oxidized alizarin and obtained "alizarinesäure," which Gerhardt (81) proved was phthalic acid. This evidence convinced Adolph Strecker, another pupil of Liebig, that alizarin was a naphthalene derivative (187), an erroneous idea which was the basis for research by Carl Glaser and Martius and Griess. Just in time Baeyer developed his zinc dust reduction process. Graebe, the assistant, and Liebermann, Baeyer's pupil, distilled alizarin with zinc dust and obtained anthracene (94, 95). They gave it the structure of phenanthrene, but their discovery ended the idea that alizarin was a naphthalene derivative. On November 18, 1868 British patent 3850 (142) was obtained by Carl Liebermann and Carl Graebe for preparing alizarin from dibromo- and dichloroanthraquinone by heating with potassium hydroxide. Graebe and Liebermann beat William Henry Perkin to the patent office by two days! They also obtained U.S. patent 95465 (96, 141). However, this process was unsatisfactory for plant production, and later Caro, Graebe, and Liebermann obtained a British patent (52), which disclosed that alizarin was formed upon fusing sodium 2-anthraquinonesulfonate with a nitrate or chlorate.

Two further examples of Graebe's work in structural organic chemistry may be mentioned. He suggested the terms ortho, meta, and para for the 1,2-, 1,3-, and 1,4-disubstitution positions on the benzene ring (91). In the same paper he proposed that naphthalene was equivalent to "two benzene rings which have two atoms of carbon in common" (91, 93).

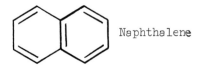

Naphthalene

This formula explains the formation of phthalic acid by oxidation, and its conversion into the anhydride shows the presence of the two adjacent or ortho-substituted carboxyl groups.

Carl Glaser. Carl Andreas Glaser was born June 27, 1841 in Kircheimbolanden at Donner's Mountain in the Rhenish Palatinate. He was the son of a physician, who died when Glaser was eight years old.

. . Courtesy M. Frosch, Badische

Figure 6.   Carl Glaser

Young Glaser graduated from the local "Progymnasium" in 1855 and from the vocational school at Kaiserlautern in 1858. From there he went to the Polytechnic school at Nuremburg and then to the Polytechnic at Munich, where he unenthusiastically studied architecture and engineering. The turning point came when Glaser attended Liebig's lectures at the University in 1862. Like Kekulé and so many others, Glaser switched to chemistry. He then went to Erlangen where, while recovering from a brief illness, he saw a copy of Kekulé's new "Lehrbuch" (*133*). This sealed Glaser's desire to be a chemist, but he had to withstand the opposition of his relatives, for they felt that there was no economic future for a chemist. Glaser soon learned that Strecker had postulated that alizarin must be derived from naphthalene (*178*). This was an erroneous assumption, but nonetheless it set Glaser to repeating the halogenation (with chlorine and bromine) of naphthalene. The study was not completed until the summer of 1864 while he was working under Strecker at Tübingen. This meticulous experimental work gained him his Ph.D. in 1864 (*85*).

Figure 7.   Kekulé with students at Bonn

At about this time Kekulé mentioned to Strecker that he was looking for an assistant. The latter immediately recommended Glaser, and told him to contact Kekulé. They met at the convention of Natural Scientists in September 1864, and Kekulé invited Glaser to come to Ghent at an annual salary of 2000 francs. On October 15 Glaser arrived to assist Kekulé. In organizing and developing the material for his lectures Kekulé took the time to write them up as they would appear later in his "Lehrbuch." In this way many new problems arose which needed experimental clarification. Glaser was an able person to carry out these investigations—so much so, that Kekulé, at the end of his long paper on the benzene theory published in the *Annalen*, wrote (*122, 183*):

I cannot conclude these communications without thanking my assistant Dr. Glaser for his valuable assistance, which he rendered in executing the described experiments.

Glaser remained at Ghent as teaching assistant until Kekulé accepted the call at Bonn University. Jean Servais Stas offered Glaser a position at the Agricultural College in Gembloux, and at the same time Kekulé offered him the position of first assistant at Bonn. While this latter position offered only 1500 marks, Glaser followed Kekulé and was established as privatdozent in the summer semester in 1868. An interesting sidelight of this period is reported by Glaser when writing about Heinrich Brunck (*87*).

We were in constant correspondence regarding the first World Exposition in Paris [1867], and I succeeded in obtaining a loan from my relatives of 200 golden florins which was enough for the trip and for a two-week stay in Paris. Our friend, Ladenburg, found cheap lodgings for us in the Latin Quarter. We were taken by Alph. Oppenheim and A. Ladenburg to a meeting of the Societe chimique where we met Ad. Wurtz, Ad. Naquet, C. Friedel, C. Lauth, A. Gautier, and others.

We frequently visited the poorly equipped laboratory at the Sorbonne, and we profoundly enjoyed the immensity of what was offered at the World's Exposition. As a member of the jury, Kekulé could point out to us items of special interest in the chemical division. For the visits to the museums, we had as our guide the art loving and art expert, Ladenburg. Our short sojourn in Paris was quite instructive, and we returned home highly satisfied.

In the spring of 1869 Kekulé told Glaser that Gustav Siegle was looking for a capable chemist for his dye factory, but Glaser did not wish to enter industry at that time. At the end of the summer semester, Glaser visited his sisters in Mannheim. Even before this Glaser had known (*87*):

that my friend Graebe, connected with the Ludwigshafen Aniline Works, was working together with Caro to make possible the large scale production of artificial alizarin from anthracene by Liebermann's process. I had been interested in the alizarin problem since 1864, and this was the reason that I went to Strecker.

Glaser further writes that while in Mannheim:

I visited Graebe in the factory, met Caro, and followed with increasing interest their process. In repeated visits I met the managing directors Friedrich Engelhorn and August Clemm.

Glaser was offered a position to work with Caro on developing the manufacture of alizarin dyes, which he accepted after his friend Brunck was offered a position as assistant to August Clemm. Glaser (87) gives some details about the aniline works as he wrote that:

[Brunck's] activities comprised a number of manufacturing processes for the production of aniline dyes, such as the purification of benzene, the separation of benzene from nitrobenzene, the manufacture of aniline, and the preparation of fuchsine. Some of the aniline dyes were reserved for A. Clemm, and others were handled by a Frenchman named Duprez. The directors, Carl Clemm and Julius Giese, were responsible for the inorganic products while Caro and I were responsible in the laboratory for the production of alizarin, induline, and the benzidine azo dyes. At that particular time these few people represented the entire chemical personnel of the Badische Anilin und Soda Fabrik.

ALIZARIN. Before the appearance of synthetic alizarin, the annual production of natural alizarin, at 100% concentration, amounted to 750,000 kilograms and was worth about 60 million German marks. It was the money crop for many farming people in the world. At Avignon stood a monument to honor the man who had benefited his countrymen by introducing madder root as a staple industry in the Department (174). This was the situation in 1868 when Graebe and Liebermann made their remarkable synthesis. Caro and Graebe began near the end of April 1869 to develop an improved process. The first improvement came when the dibromoanthraquinone process was discarded for one from anthraquinonesulfonic acid. The work itself offered great difficulties. The starting material, anthracene, amounted to only 0.4% of coal and had to be purified by a laborious procedure. As a by-product to this work, Glaser discovered carbazole and phenanthrene in crude anthracene during 1872. However, the main troubles with the alizarine production were solved by the end of 1870. The fusion of anthraquinonesulfonic acid with alkali was improved when Glaser developed his "druckschmelze" or pressure fusion process in which potassium chlorate was added to the melt (taken from Kekulé's phenol work) in a specially constructed and agitated pot. This gave "a perfect product, an alizarin which could not be produced by the competition" (4).

In the first year the production of alizarin amounted to only about 2000 kg. of 100% dye. However, it continued to rise until 100,000 kg. were made in 1875. The yearly consumption of the synthetic material had risen to 2 million kg. by 1902.

In 1879 Glaser and Brunck were made associate directors of Badische, and in 1884 they, along with Hanser, took over the business management.

Glaser was a director from 1884 to 1895, was on the board of directors from 1895 to 1918, and served as chairman for the last seven years. He died in Heidelberg on July 25, 1935 (*4, 33, 83, 84, 87, 172*).

Heinrich Brunck. Carl Glaser wrote (*87*):
I spent the end of August 1865, a part of my autumn vacation, with relatives at Kirchheimbolanden, my native city. My chemical studies had been frowned upon. (They were of the opinion that a student without means should be interested in earning money, since chemistry is an art which produces no bread). But to the surprise of all my acquaintances, I succeeded in becoming assistant to Professor Kekulé at Ghent, with an annual income of 2000 francs, after I passed my doctor's examination. I rose considerably through this renown, so that the attention of an older brother, Ulrich Brunck, a farmer, was focused upon me. He came to see me to ask about Heinrich's wish to study chemistry. Heinrich had some chemical training at the Zurich Polytechnicum, but then he went to Tübingen and became a member of a student corps (fraternity), Suevia. Here he enjoyed himself but did not study. The parents and older brothers were afraid that when Heinrich attended another German Technical College he would again become involved in the student fraternity.

Courtesy M. Frosch, Badische

Figure 8.   Heinrich Brunck

I proposed to take the young man along with me to Ghent. At this University he could learn a lot, and also learn how to speak French, but it would be impossible to continue in the student fraternity. This proposition was immediately accepted by the brother, and he reported back to the Brunck family, who were located in Winterborn. Heinrich and I spent one week together and formed a friendship which could only be separated by death.

The Brunck family lived in the small village of Winterborn, high on a plateau of the Alsenztal of the Bavarian Palatinate. Heinrich was born here on March 26, 1847, the youngest of five brothers and three sisters. On October 20, 1965 Glaser met Heinrich Brunck at Saarbrücken, and they travelled through Luxembourg to Ghent. Since Kekulé was away, Brunck was assigned to Theodor Swarts. Glaser reported on the impression Brunck made (87):

The tall, slim, young fellow with curly black hair, bright brown eyes, fresh complexion, and a charming winning manner, at only eighteen and one-half years of age won the hearts of the small circle of Germans who had been attracted to the Ghent laboratory by Kekulé's reputation. I mention a few of these coming chemists, such as W. Körner, A. Wichelhaus, A. Ladenburg, C. Leverkus, Ad. Mayer, Esch, Behrend, and Semmel.

Courtesy B. Helferich, Bonn

Figure 9. Kekulé with pupils at Ghent. Standing: August Mayer, Körner, Esch, Semmel, Behrend, A. Ladenburg. Sitting: T. Swarts, Kekulé, Carl Glaser

There existed an active spiritual life in this small circle during the day. With exception of a pause for lunch, the work went on swiftly. The meals were taken together at the Hotel de Vienne. On some special occasion, as when a barrel of Munich beer had found its way to the Hotel de Vienne, there would appear Kekulé. He was then at the height of his fame. This was a festive occasion for us as we clung with great admiration to the master. Even behind the beer table he was forceful as he told with great vivacity of his experiences with Liebig, Williamson, Dumas, and others.

At first Brunck was assigned to general synthetic work until Körner gave him the problem of phenol substitution. His first project was to brominate the two isomers of nitrophenol and their mono- and disubstituted compounds along with a few salts of these derivatives. Körner was able to watch Brunck's progress since their work benches were in the same laboratory, and his experimental technique greatly aided Brunck. This work formed the basis for Brunck's Ph.D. dissertation at Tübingen in 1867. He spent the next semester listening to Johannes Wislicenus' lectures at Zurich University, and then went with Glaser to the first World Exposition in Paris. Upon returning home Brunck was employed by de Haen for his firm at List, near Hannover, and remained there until he joined Badische.

At Badische, Brunck at first was concerned with the production of fuchsine and its intermediates and later with the purification of anthracene, which led to the alizarine process and other alizarin derivatives. Meanwhile, Glaser prepared an alizarine derivative which dyed mordanted cotton a pure blue, and Brunck took over the development of the manufacturing process for Alizarin Blue (97). Brunck found that by

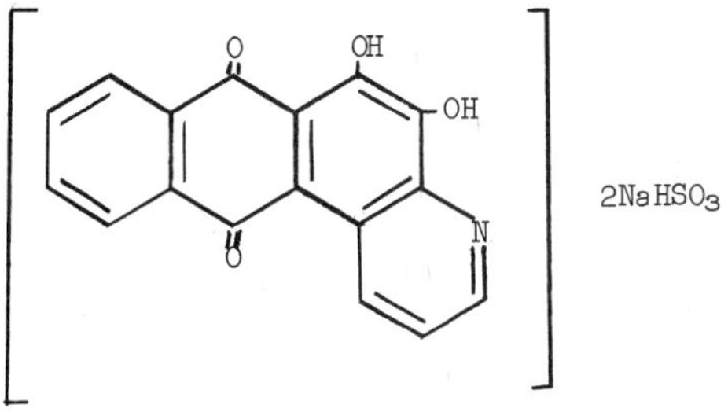

Alizarine Blue S

heating the dye with sodium bisulfite, a soluble form was obtained called Alizarin Blue S. This dye (*12, 41, 43*) proved valuable for producing indigo-shade dyeings but was eclipsed by the subsequent appearance of synthetic indigo.

INDIGO. The story of indigo is undoubtedly the story of the first dye in common use. Long before man could clothe himself in "purple and fine linen" he painted his body in the West with woad extract and in the East with indigo extract. Marco Polo is said to have brought the dye to Europe from his journey to India about 1300. Indigo became so popular that its use was equal to that of all other dyes combined. Small wonder then that dye companies like Badische (54) spent much money and effort in an unsuccessful attempt to use Baeyer's synthetic methods (26). Finally, they came to realize that even if it were developed, there was not enough benzene and toluene to produce the quantities of indigo needed to supply the world's consumption—estimated then at about 11 million pounds. However, naphthalene was available in sufficient quantity, and Badische purchased the patented process of K. Heumann (108) and others. A practical, economical manufacturing process was finally achieved (144). The following series of reactions summarize the tortuous but successful route:

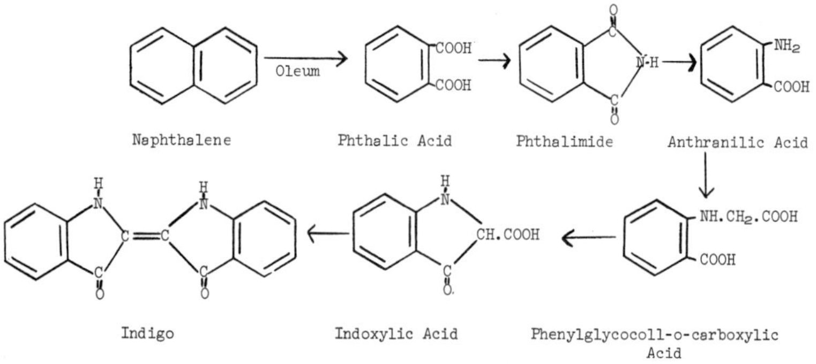

| Naphthalene | Phthalic Acid | Phthalimide | Anthranilic Acid |

| Indigo | Indoxylic Acid | Phenylglycocoll-o-carboxylic Acid |

Brunck (42, 44) and Rudolf Knietsch under the aegis of Glaser, were finally able to see synthetic indigo make its entry into the world market. This product all but doomed native indigo in the same way that synthetic alizarine had eliminated the growing of madder root.

Brunck, like Glaser, served as manager, president, and chairman of the board of directors of Badische. Brunck was general manager when a colorful brochure was prepared for distribution at the International Exposition at Paris in 1900 (10). This was a far cry from his visit in the company of Glaser to the 1867 exposition. Heinrich Brunck died December 4, 1911 (4, 38, 45, 46, 48, 61, 87, 118, 144, 151, 152, 179).

Heinrich Caro. Heinrich Caro's work was without doubt the first chapter in the early dye history. He was a genius endowed with talents —both good and bad—whose name will be remembered as long as a history of synthetic dyes remains. Naturally, he wrote the first history, "On the Development of the Coal-tar Color Industry" (53). Caro was born at Posen, Poland, on the banks of the Warta River, on February 13, 1834. He grew up in Berlin, where he attended the Gymnasium, went to the University for pure science, and finally studied at the "Gewerbeakademie" from 1852 to 1855. Here he learned textile dyeing and printing, and in 1855 he obtained a job in a dyehouse at Mulheim-on-the-Ruhr. In

Figure 10.   Henrich Caro

spite of the secrecy which abounded in the Alsatian dyehouses, Caro
quickly learned what was going on. He soon saw the absurdity of the
ancient belief that dyeing in the winter was impossible and managed to
keep production going regardless of the season. It did not take management long to realize the abilities of this young man. Caro was sent to
England in 1857 to study the advances of the textile industry, to buy
machinery, and to learn more about color printing on cotton. He was so
impressed with what he saw that he returned.

In 1859 Caro arrived in Manchester, and one of his part-time jobs
was conducting chemical analyses for Roberts, Dale, and Co. Its founder,
John Dale (1815–1889) was an enterprising young man, who saw that
association with Caro and other notable chemists as Martius, Schad,
Leonardt, and Keopp would greatly advance the synthetic dye industry.
Dale had come to Manchester to assist a Mr. Ansell, a druggist and chemist. Since Ansell was a personal friend of John Dalton, it was not long
before Dale became Dalton's pupil. In 1852 he and Thomas founded the
firm of Roberts, Dale, and Co. at Cornbrook.

In 1860–61 Dale and Caro discovered that they could use a mixture
of copper sulfate and alkali chloride to replace the expensive potassium
dichromate in synthesizing Perkin's Mauve and that chalk or some other
cheap alkali could be used for the final neutralization. Many new dyes,

patented under the names of Caro, Martius, and Dale, followed quickly. Manchester or Bismark brown (51, 59) was synthesized in 1864; Martius yellow (Martius' first dye) (2,4-dinitro-1-naphthol) was made in 1864; Aniline yellow, later proved to be the oxalate of aminoazobenzene, was first marketed by Dale's company; the spirit-insoluble indulines in 1863. Dale was the first to manufacture picric acid from phenol rather than from Australian gum.

Contributions such as these helped establish England's position in the synthetic dye industry. While the Literary and Philosophical Society of Manchester is best remembered for Joule, the physicist, and Dalton, the theoretical chemist, it can also be proud of John Dale, industrial technologist (57). Had there been more of his breed, England might have remained foremost in the dye industry.

While working for Dale, Caro instituted a life-long pattern of persuading management to use scientific consultants. Among those at Dale were Peter Griess and Carl Schorlemmer. Before long Caro became a partner in the company, but in 1866 he resigned and returned to Germany. He first went to Heidelberg for study and experiment, and on November 1, 1868, he joined Badische (4, 39, 172). Until 1899 he was research director, and from 1884 to 1890 he was director of the company. Naphthol yellow (13), alizarine, methylene blue, fast red (55), Auramine, and many other colors were either discovered or developed by him.

His artistic temperament made it difficult for conservative chemists to work with him on a day-to-day basis. For example, Müller writes (4):

It must not have been an easy task for Glaser to work with Caro, who was of an entirely different make up and nature and who found it hard to stick to a given problem.

Along the same line, E. F. Ehrhardt states (67):

I went to Ludwigshafen and was engaged as a chemist under Caro. When after six months' provisional engagement the definite agreement was to be signed, I asked for a salary of 3000 marks per annum as against the regular 2400 marks and gave as the reason for this 30 (English) pounds extra that for me the engagement meant living abroad, so that something ought to be paid by way of consolation for living in exile. Dr. Brunck who was dealing with me in the matter, snorted at this idea and said in other words, "If you know a better 'ole, go to it," but he added, "we will give you the extra 30 (English) pounds, not for living with us here, but you have got to work under Caro, no one can get on with him, and we have noticed in the last six months that you have got on with him. We will give you the extra 30 (English) pounds per annum for that."

However, Ehrhardt (67) added:

It was, and I suppose still is, a social custom over there for the new recruits on the staff to call on the older man, and these calls were paid between the hours of 11 and 1 on Sunday morning. I promptly paid my duty call on Dr. Caro. Instead of letting me go at the end of the regulation 10 minutes he kept me until he was fetched to dinner and insisted on taking me with him. He kept me the whole afternoon. We went for a walk and returned. He kept me for the evening meal, and for the whole of the evening until past midnight filled the time with practically a monologue on the history of the dye industry and his experiences in connection with it. He made this most interesting to me, and he thoroughly enjoyed talking in this way himself and pressed me to come again, and repeatedly my morning calls ended only at midnight.

It was not long after this that Caro wrote at least three histories of synthetic dyes. His first, "To the memory of Peter Griess" (49), was undoubtedly the first survey of diazo and azo compounds. In 1892 he published "On the Development of the Coal Tar Dye Industry" (53), a 150-page story of synthetic dyes which he was eminently qualified to write. Finally, in 1904 Caro wrote "On the Development of the Chemical Industry at Mannheim-Ludwigshafen on Rhine" (48), an account of the growing Badische companies as related to the evolving story of the chemical and dye industries.

A total of 26 dyes, covering the direct, basic, mordant, and solvent classes, were discovered by Caro. After retirement, he continued his private research until his death on September 11, 1910. One discovery from this period (1898) was "Caro's Acid," $HO.O.SO_3H$, a strong oxidizing agent (33, 46, 62).

August Bernthsen. Heinrich August Bernthsen was born at Krefeld on August 29, 1855. He was a pupil of and later spent three and one-half years as assistant to Kekulé. Anschütz says that Bernthsen was the instructor-assistant in one of the analytical laboratories and had charge of the lecture assistants. He also says that Bernthsen's experimental technique was excellent. On Kekulé's advice Bernthsen left in the spring of 1879 to become privatdocent at Heidelberg, where he had been Bunsen's pupil in the summer of 1874. Bernthsen became a professor and remained at Heidelberg until 1887. Badische had built a new central laboratory, and Bernthsen was invited to replace Caro who had become too enmeshed in patent affairs to continue as its director. Caro continued to look after the aniline dyes, azo dyes, and patent affairs until his resignation in 1889 when Bernthsen took over entirely. For 30 years he was an active researcher, and near the close of this period he was a patent attorney. Bernthsen stayed until 1918, when he resigned to return to Heidelberg where he was made an honorary professor at the University in 1920.

Bernthsen had already gained a reputation through his fundamental work on the constitution of methylene blue before joining Badische. A total of 5 dyes listed in the "Colour Index" are credited all or in part to Bernthsen—i.e., Lauth's violet, methylene azure, methylene blue, toluidine blue, and rhodamine 6G. However, these are a poor indication

Courtesy M. Frosch, Badische

Figure 11.   August Bernthsen

of the value of his work at Badische, where he became a member of the board of directors. Bernthsen may be recalled by several of us here as the author of the Bernthsen-Sudborough: "A Textbook of Organic Chemistry." He died at Heidelberg on November 26, 1931 (*1, 34, 115, 119, 149*).

Gustav Schultz. Gustav Theodor August Otto Schultz, born December 15, 1851 in the Westphalian town of Finkenstein, was a pupil of Graebe at Königsberg where his doctoral thesis was "On Diphenyl" (*167*). Graebe had suggested that Schultz work on the constitution of benzidine, and he came so proficient on the reaction that he became known as "Diphenyl Schultz." In 1876 he came to Bonn as Kekulé's assistant and remained until 1877. During the winter semester of 1877–78 he became Fittig's teaching assistant at Strassburg. There he was married and soon afterwards became a lecturer in organic chemistry until 1882, when he left teaching to become research director of Aktien Gesellschaft für Anilin fabriken (Agfa) in Berlin. He remained in Berlin for 12 years—a productive period both as researcher and writer.

The "Colour Index" lists 24 dyes of which he was either the inventor or co-inventor, including a variety of acid dyes (Erika reds) and direct dyes (Hessian bordeaux, purples, and violet). The most important were probably brilliant yellow and benzopurpurin 4B. His assistant, F. Bender, is credited with methylating brilliant yellow to obtain the valuable chrysophenine. Schultz's literary efforts were splendid. While at Bonn he had joined with Anschütz in prodding Kekulé to continue his "Lehrbuch." They started writing sections, either alone or with Kekulé, until

Courtesy Verlag Chemie

Figure 12.   Gustav Schultz

the first section of the fourth volume appeared in 1887; the project was then dropped. In 1882 Schultz published his "Chemistry of Coal Tar, with Special Consideration of the Synthetic Organic Dyes." By 1887 he had joined forces with Paul Julius in compiling their first edition "A Tabular Survey of Synthetic Dyes," published in *Chemische Industrie* (*169*) and partly reprinted in the *Journal of the Society of Chemical Industry* (*170*). The next edition appeared in book form. Arthur G. Green came out with the first English translation in 1894, a volume which initiated the two editions of the "Colour Index." Gustav Schultz was on the committee in Berlin which arranged the "Celebration of the German Chemical Society to honor August Kekulé." It was at this glorious celebration in 1890 that the portrait of Kekulé by Angeli was presented to him by friends in the dye industry. Schultz reported the affair in a 49-page account in *Berichte* (*168*). In 1895 he went to work for a Basel dye plant, but the next year became professor and head of the Chemical Technology Department at Munich's Technische Hochschule. Here for the next 30 years he turned out chemists for the dye and allied industries. Schultz died following a long illness on April 21, 1928 (*1, 171*).

Paul Julius. Paul Julius was born October 1, 1862 at Liesing near Vienna, the son of a chemist. He studied at the Technische Hochschule in Vienna as the pupil of Benedikt and later became assistant to Weselsky and Skraup. In the autumn of 1885 he went to Basel to further his work

Courtesy Verlag Chemie

Figure 13.   Paul Julius

in dye chemistry, where he came under the influence of Nietzki and Noelting. By March 11, 1886 he had finished two outstanding works: "On a New Diamidodinaphthyl" (*121*) and "On the Composition of Magdalen Red" (*120*). This was enough evidence for Bernthsen and Caro to bring him to Badische in June 1887, where he worked for 27 years and succeeded Bernthsen as director of the laboratory. He assisted Jacobsen in editing "Die Chemische Industrie," and by 1887 he had written "The Synthetic Organic Dyes." Soon he collaborated with Schultz to prepare the first tables of organic coloring matters: "Tabular Survey of Synthetic Organic Dyes" (*169*). Julius prepared and identified many dye intermediates—e.g., J-Acid (2-amino-6-naphthol-7-sulfonic acid) and M-acid (1-amino-5-naphthol-7-sulfonic acid) and was active in the field of substantive disazo dyes. Bernthsen and Julius patented six of these disazo dyes, which are still listed in "Colour Index"—Oxamine violet and blues. He is credited with seven other dyes, which include a mordant, acid, basic, sulfur, and two vat dyes. While not one of Kekulé's pupils, he lived and worked among them and was considered a friend. Julius died.unexpectedly in Heidelberg on January 9, 1931, in his 68th year (*143, 170*).

Wilhelm Körner. Wilhelm Körner was born in Cassel on April 20, 1839. After finishing school at the local polytechnic, he went to Giessen to study chemistry. Liebig had gone, but the influence of Will, Kopp, and Engelbach stimulated Körner, and after graduating in 1860, he stayed on as an assistant for three years; then he went to Ghent to work under Kekulé. He remained there until 1867, except for part of 1865 when he went to London to serve William Odling as assistant at St. Bartholomew's Hospital, where Kekulé had assisted John Stenhouse earlier. Körner was closely associated with Kekulé during the development of the benzene theory, and it was no accident that a paper by Körner followed Kekulé's paper in *Annalen* (*137*). Körner realized the practical consequences which would result from establishing the benzene theory on a sound experimental basis. His own efforts were collected in his great memoir: "Researches on Isomerism amongst the So-Called

Courtesy Chemical Society, London

Figure 14. Wilhelm Körner

Aromatic Substances Containing Six Atoms of Carbon" (*138*), published in 1874. Armstrong wrote (*5*):

A vast body of unimpeachable laboratory work was therein described, which solved the problems of relative position in a most masterly manner and for all time, insofar as these can be expressed against a regular hexagon symbol.

Körner left Ghent in 1867 to go south for his health. He reached Cannizzaro's laboratory at Palermo and stayed there until 1870, when he was elected to the chair of organic chemistry at the newly opened School of Agriculture at Milan. Körner resided in Italy the rest of his life. He loved and was loved by the Italians to the extent that he replaced his Christian name, Wilhelm, with Guglielmo in his personal and professional relations. He died at Milan on the evening of March 29, 1925 (*6, 7, 56*).

Carl Duisberg. Carl Duisberg was born September 29, 1861 in Barmen, a small textile center near Elberfeld. These two towns later merged to form Wupperthal. The only son of Johann Karl Duisberg, a ribbon maker, young Duisberg rejected the idea of entering the family business in favor of becoming a chemist. He went to Göttingen in 1879, where he first came under Paul Jannasch and Hans Hübner, a pupil

Courtesy Verlag Chemie

Figure 15.   Carl Duisberg

of Kekulé and successor to Friedrich Wöhler. Since he could not receive a degree there (owing to regulations) he transferred to Jena. Anton Geuther, a student of Wöhler's and a disciple of the Wöhler-Liebig school of chemical instruction, taught Duisberg the fundamentals of chemistry. He progressed so quickly that he received his Ph.D. at 20 with a thesis on ethyl acetoacetate (*63*).

Duisberg later wrote a memorial to his old teacher, in which he related the amusing story of his final departure from Geuther's laboratory (*64*). Armstrong described the departure as follows (*8*):

Anxious to be no longer a burden on his parents, after four years study at the University, even before he had secured his Doctor's degree, he applied for a post as chemist to the food analysis department at Krefeld. Geuther protested and paid him the compliment of saying that he was too good to be an analyst. He proposed that he should become his private assistant at a salary of 80 marks with free lodging in an attic, subject to the condition that he should remain with him until he obtained a suitable post in industry. Evidently Geuther already appreciated his ability. On seeking a post at the end of the summer term, he found the objection raised everywhere that he had not yet been through his military service. So he told Geuther that he proposed to leave him in autumn, to join up with an infantry regiment in Munich. This put Geuther in a towering rage; he insisted that Duisberg had promised to remain with him until he could enter industry. All argument was in vain. The dispute had taken place angrily in the laboratory at Duisberg's bench, where he was busy cleaning a large spherical flask. Waving the flask in the professor's face he followed him to the door; finally, as Geuther persisted in his contention, Duisberg dashed the flask violently at his feet, so that it broke into a thousand pieces. I can well picture the young gascon thus bringing down the curtain.

Although an innate quality, there is little doubt that the severity of his training under Geuther was of extreme value in developing the wonderful technical sense which has been the cause of his great success as an industrial leader.

Duisberg proceeded to Munich, where he served for one year in the army and spent his evenings studying under Baeyer at the University. This labor resulted in a joint paper with Hans v. Pechmann on the synthesis of coumarins from phenol and ethyl acetoacetate (*150*). Duisberg was hired by Carl Rumpff of the Farbenfabriken vorm Friedrich Bayer and Co. A.G. at Elberfeld on September 29, 1883. He was immediately sent to work under Rudolph Fittig at the University of Strassburg to evaluate P. J. Meyer's synthesis of isatin derivatives for producing indigo. He returned to Bayer in 1884 and immediately began to synthesize dyes using homologues and analogues of benzidine—e.g., *o*-tolidine and benzidine sulfonedisulfonic acid. In 1885 came benzoazurine, benzopurpurine B,4B and 6B, and sulfonazurin. Duisberg's azo blue, made from tetrazotized *o*-toluidine and coupled with two moles of Nevile-Winther's acid, was reported to be the first blue azo dye. A series of dyes followed, mostly made in collaboration with others, until azocochenille in 1892. In 1900 Carl Hermann Wichelhaus listed Duisberg's technical accomplishments as five intermediates, 26 dyes, and one pharmaceutical—phenacetin (*29, 65, 109*).

These valuable discoveries resulted in Duisberg's being appointed as a director of Bayer in 1888; from then on his administrative talents were given full opportunity, especially since he married Johanna Seebohm, a relative of his employer, Carl Rumpff. Duisberg fulfilled everyone's belief in him, to the benefit of Bayer and the whole German dye industry.

In 1891 Duisberg completed a new laboratory at Elberfeld (29), and realizing the need of a good library, he bought the personal libraries of deceased chemists. The first large collection of about 7000 volumes belonged to August Kekulé and was purchased for 28,000 marks (32, 66). The company had about 3000 volumes at the time of this purchase (146); hence, the company library became known as the Kekulé Library. The following year Duisberg acquired the libraries of Victor Meyer and Henry E. Roscoe.

Carl Duisberg died at Leverkusen on March 19, 1935 (9, 155, 161, 166).

Otto N. Witt. Otto Nikolaus Witt was born March 31, 1835 in St. Petersburg. Johannes Niklas Witt, his father, was originally from the duchy of Holstein, and was teaching science at the state technical institute. The family moved to Munich when Witt was 11 and to Zurich

Courtesy Technische Universität, Berlin

Figure 16.   Otto N. Witt

Figure 17.   A. Wilhelm Hofmann

two years later, where he attended the gymnasium, the "Industriesschule," and entered the Polytechnicium in October, 1871. Here he was influenced by J. Wislicenus and E. Kopp, and worked on *m*-dichlorobenzene and diphenylnitrosamine. On July 13, 1875 Witt graduated "promvierte" at the University of Zurich and proceeded to England where he became a chemist for Williams, Thomas, and Dower in Brentford, near London. His first discovery, made independently by Caro, was chrysoidine (1875) (*185*). This was followed by other yellow to orange-yellow azo dyes for wool, one being tropaeoline. But back a moment to the confusion surrounding chrysoidine.

Greiss wrote to Hofmann on February 20, 1877:

Chrysoidine was discovered by Mr. Caro in Mannheim, and independently from him, by Mr. Witt in London. It was brought into commerce for the first time by his firm, Williams, Thomas, and Dower. In the South Kensington Exhibition a beautiful sample of chrysoidine was exhibited last summer by Mr. Witt under his name.

Both Witt and Caro met at the exhibition and decided to say or do nothing for the time being about the dye. Early in 1875 Griess had known that he and Caro were both investigating chrysoidine. Griess had concluded:

The formation of chrysoidine dye is entirely independent of the nature of the diazo compound but is dependent upon the nature of the constitution of diamino compounds and, in this way, the two amino groups are in the 1,3- or meta position according to Kekulé's theory.

Hofmann, in the meantime, finding no public record, published his own investigation of chrysoidine. Witt objected since he had felt that the description in the exhibition catalogue was sufficient. In response, Hofmann wrote (114):

Let us call a spade a spade, as Mr. Griess has done this with true spirit. It is obvious that Mr. Witt had "business reason" for not disclosing his beautiful discovery to the world. Who could blame him for this? Why should a chemist not utilize the fruit of his brain labor just like an author or artist? As far as the road to follow is concerned there can be different viewpoints. If a chemist, however, decides to keep the nature of his discovery secret while he markets his product, so that everybody can buy it, he must not be surprised when the secret is only of ephemeral duration. The time of the "Arcanists" is over. Whosoever, in the last quarter of the nineteenth century, will give his colleagues a chemical puzzle must be prepared to see it solved sooner or later.

Young Witt learned his lesson well and eventually published a total of 108 papers—the first in 1873 and the last in 1915.

While in England Witt also discovered safranine, induline, azophenine, and toluylene blue and red. He left England to join L. Casella in Frankfurt-am-Main in 1879, but again he transferred to the chemical institute at Mülhausen about 1880. Here he synthesized neutral red, neutral violet, indophenol, and other dyes. From 1882 to 1885 he was a chemist for Verein Chemischer Fabriken in Mannheim, and his investigations resulted in nitroso derivatives of aromatic diamines, eurhodine, products from the sulfonation of naphthylamine, and 1,4-naphtholsulfonic acid (Nevile-Winther's acid). The "Colour Index" lists 16 dyes as being discovered by Witt.

In 1876, while in England, Witt started working on his color theory, coining such words as chromophor—e.g., $-NO_2$ and $-N{=}N-$, and chromogen—e.g., nitroaniline and anthraquinone, to describe those groups or side chains which contribute to or enhance the formation of color in a dye molecule (184).

In 1885 he went to Berlin, where he graduated in 1886 from the Charlottenburg Technische Hochschule. He became a professor here in 1891 and Rector in 1897. Witt died in Charlottenburg (Berlin) on March 23, 1915 (60, 186).

I cannot leave Witt without calling attention to his humorous literary effort with Emil Jacobsen. During 1886 Kekulé was serving as president

Fig. 1.   Fig. 2.

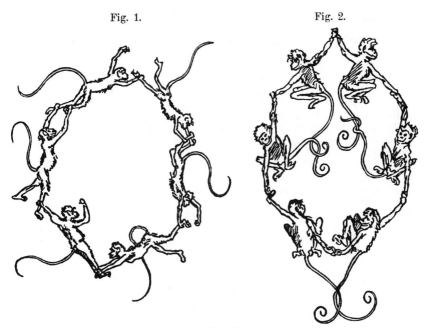

Figure 18.   Monkey benzene rings

of the German Chemical Society. In September a delightful little spoofing pamphlet appeared, entitled, "A report of the Thirsty Chemical Society, Unheard of Volume, No. 20" (issued September 20). One example concerned a paper by "F. W. Findig: Contribution to the Constitution of Benzene" in which monkeys replace carbons in Kekulé's benzene hexagon (*182*).

Dedication and Acknowledgements

I would like to dedicate this paper to the memory of that adept and selfless dye historian, Wilfred H. Cliffe of Manchester. He would have enjoyed helping me, while spicing his aid with humor, but his long and fatal illness restrained me from telling him of this project.

Among those whose help must be gratefully acknowledged are C. Brauer, Nationalgalerie, Berlin; Ralph Disler, Tennessee Eastman; Sidney M. Edelstein, New York; Frederick R. Greenbaum, Margate City, N. J.; B. Helferich, Bonn; Helmut Möhring, Bayer, Leverkusen; I. Pohle, Adolf von Baeyer Bibliothek, Frankfurt; C. Schöpf Darmstadt, R. J. Smith, I.C.I., Manchester; Maria Frosch, Badische, Mannheim; Luise Michel-Glaser, daughter of Carl Glaser, Mannheim; Ruth S. Henley, Tennessee Eastman; Stella W. Wilcox.

Literature Cited

(1) Anschütz, R., "August Kekulé," Vol. 1, pp. 608-9, Verlag Chemie, Berlin, 1929.
(2) *Ibid.*, Vol. 2, p. 803.
(3) Anschütz, R., *Ann.* **294**, 232 (1897); **306**, 1 (1899).
(4) Anschütz, R., Müller, C., *Angew. Chem.* **40**, 273 (1927).
(5) Armstrong, H. E., *J. Chem. Soc.* **1876**, 204.
(6) Armstrong, H. E., *J. Chem. Soc.* **1887**, 258, 583.
(7) Armstrong, H. E., *J. Soc. Chem. Ind.* **48**, 914 (1929).
(8) Armstrong, H. E., *J. Soc. Chem. Ind.* **50**, 260 (1931).
(9) Armstrong, H. E., *Nature* **135**, 1021 (1935).
(10) Badische Anilin- und Soda-fabrik, "The BASF Digest," Vol. 11, Ludwigshafen am Rhein, 1964.
(11) BASF, German Patent **34294** (June 18, 1885).
(12) BASF, German Patent **17695** (August 14, 1881).
(13) BASF, German Patent **10785** (Dec. 28, 1879).
(14) BASF, German Patent **1886** (Dec. 15, 1877).
(15) Baeyer, A., *Ann.* **103**, 178 (1857).
(16) Baeyer, A., *Ann.* **103**, 181 (1857).
(17) Baeyer, A., *Ann.* **105**, 265 (1858); **107**, 257 (1858).
(18) Baeyer, A., Emmerling, A., *Chem. Ber.* **2**, 679 (1869).
(19) Baeyer, A., Caro, H., *Chem. Ber.* **10**, 692, 1262 (1877).
(20) Baeyer, A., Caro, H., *Chem. Ber.* **10**, 818 (1877).
(21) Baeyer, A., *Chem. Ber.* **11**, 582 (1878).
(22) Baeyer, A., *Chem. Ber.* **11**, 1228, 1296 (1878).
(23) Baeyer, A., *Chem. Ber.* **13**, 2254 (1880).
(24) Baeyer, A., Landsberg, L., *Chem. Ber.* **15**, 57 (1882).
(25) Baeyer, A., *Chem. Ber.* **16**, 2188, 2204 (1883).
(26) Baeyer, A., *Chem. Ber.* **33**, sonderheft, li (1900).
(27) Barth, L., Senhofer, C., *Chem. Ber.* **8**, 1477 (1875).
(28) Barth, L., Senhofer, C., *Chem. Ber.* **9**, 969 (1876).
(29) Bayer, F. and Co., "Abhandlung Vortgäge und Reden aus den Jahren 1882-1921 von Carl Duisberg," Verlag Chemie, Berlin, 1923.
(30) Bayer, F. and Co., British Patents **9162** (June 19, 1884), **9606** (July 1, 1884).
(31) Bayer, F. and Co., German Patent **31658** (June 14, 1884).
(32) Beer, J. J., "The Emergence of the German Dye Industry," University of Illinois Press, Chicago, 1959.
(33) Bernthsen, A., *Angew. Chem.* **24**, 1059 (1911).
(34) Bernthsen, A., *Angew Chem.* **42**, 382 (1929).
(35) Bernthsen, A., *Angew. Chem.* **43**, 719 (1930).
(36) Bernthsen, A., *Chem. Ber.* **16**, 1025, 2896 (1883).
(37) Bernthsen, A., *Chem. Z.* **34**, (113), 1001 (1910).
(38) Bernthsen, A., *Chem. Z.* **35**, 1385 (1911).
(39) Bernthsen, A., *J. Soc. Chem. Ind.* **29**, 1143 (1910).
(40) Blomstrand, C. W., *Chem. Ber.* **8**, 51 (1875).
(41) Brunck, H., Graebe, C., *Chem. Ber.* **15**, 1783 (1882).
(42) Brunck, H., *Chem. Ber.* **33**, lxxi (1900).
(43) Brunck, H., Graebe, C., *J. Soc. Chem. Ind.* **1**, 399 (1882).
(44) Brunck, H., *J. Soc. Chem. Ind.* **20**, 239 (1901).
(45) Brunck, H., *J. Soc. Chem. Ind.* **31**, 15 (1912).
(46) Bugge, G., "Das Buch der Grossen Chemiker," 2 vols., Verlag Chemie, Berlin, 1961.
(47) Butlerov, A. M., *Z. Chem.* **1863**, 500.
(48) Caro, H., *Angew. Chem.* **17**, 1343 (1904).

(49) Caro, H., *Chem. Ber.* **24**, (1891).
(50) Caro, H., British Patent **3751** (Oct. 9, 1877).
(51) Caro, H., Griess, P., *Z. Chem.* **3**, 278 (1867).
(52) Caro, H., Graebe, C., Lieberman, C., British Patent **1936** (June 25, 1869).
(53) Caro, H., *Chem. Ber.* **25**, 955 (1892).
(54) Caro, H., *J. Soc. Chem. Ind.* **1**, 276 (1882).
(55) Caro, H., U.S. Patent **204,799** (June 11, 1878).
(56) Cohen, J. B., *J. Chem. Soc.* **1925**, 2975.
(57) Dale, J., *J. Soc. Chem. Ind.* **8**, 528 (1889) [obituary].
(58) Dale, J., Caro, H., British Patent **3307** (Dec. 31, 1863).
(59) Dale, J., Caro, H., Martius, C., British Patent **2785** (Nov. 9, 1864).
(60) D'Ans, J., *Chem. Z.* **77**, 279 (1953).
(61) Davis, W. A., *J. Soc. Chem. Ind.* **43**, 266 (1924).
(62) Duisberg, C., *Angew. Chem.* **24**, 1057 (1911).
(63) Duisberg, C., *Chem. Ber.* **15**, 1378 (1882).
(64) Duisberg, C., *Chem. Ber.* **63A** 145 (1930).
(65) Duisberg, C., *Chem. Ber.* **68A** 119 (1935) [obituary].
(66) Duisberg, C., "Meine Lebenserinnerungen," Philipp Reclam, Leipzig, 1933.
(67) Ehrhardt, E. F., *J. Soc. Chem. Ind.* **43**, 561 (1924).
(68) Engler, C., Emmerling, A., *Chem. Ber.* **3**, 885 (1870).
(69) Erlenmeyer, E., *Chem. Ber.* **7**, 1110 (1874).
(70) Erlenmeyer, E., *Z. Chem.* **1861**, 176.
(71) Erlenmeyer, E., *Z. Chem.* 678 **1863**, 678. ·
(72) Fischer, E., *Ann.* **190**, 67 (1878).
(73) Fischer, E., Fischer, O., *Ann.* **194**, 242 (1878).
(74) Fischer, E., *Chem. Ber.* **8**, 589, 1005 (1875).
(75) Fischer, E., Fischer, O., *Chem. Ber.* **11**, 1079 (1878).
(76) Fittig, R., *Ann.* **124**, 284 (1862).
(77) Frank, E., U.S. Patents **329,638** and **329,639** (Nov. 3, 1885).
(78) Frank, E., U.S. Patent **401,024** (April 9, 1889).
(79) Friedlaender, R., *Naturwiss.* **3**, 573 (1915).
(80) Friswell, R. J., Green, A. G., *J. Chem. Soc.* **1885**, 917.
(81) Gerhardt, C., *Compt. Rend.* **5**, 222 (1849).
(82) Gerhardt, C., *Rev. Sci.* **10**, 210 (1841).
(83) Glaser, C., *Angew. Chem.* **44**, 525 (1931).
(84) Glaser, C., *Angew. Chem.* **44**, 556 (1931).
(85) Glaser, C., *Ann.* **135**, 40 (1865).
(86) Glaser, C., *Ann.* **142**, 364 (1867).
(87) Glaser, C., *Chem. Ber.* **46**, 353, 1647 (1913).
(88) Glaser, C., *Chem. Ind. (Dusseldorf)* **1895**, 397.
(89) Glaser, C., *Z. Chem.* **2**, 308 (1866).
(90) Graebe, C., *Angew. Chem.* **28**, 433 (1915).
(91) Graebe, C., *Ann.* **149**, 1 (1869).
(92) Graebe, C., Glaser, C., *Ann.* **167**, 131 (1873).
(93) Graebe, C., *Chem. Ber.* **1**, 36 (1868).
(94) Graebe, C., Liebermann, C., *Chem. Ber.* **1**, 49 (1868).
(95) Graebe, C., Liebermann, C., *Chem. Ber.* **1**, 104 (1868).
(96) Graebe, C., Liebermann, C., *Chem. Ber.* **2**, 332 (1869).
(97) Graebe, C., Brunck, H., *Chem. Ber.* **11**, 522 (1878).
(98) Graebe, C., *Chem. Ber.* **41**, 4805 (1908).
(99) Green, A. G., "A Systematic Survey of the Organic Colouring Matters," pp. 76-77, Macmillan, London, 1908.
(100) Griess, P., *Ann.* **106**, 123 (1858).
(101) Griess, P., *Ann.* **113**, 334 (1860).

(102) Griess, P., *Ann., suppl.* **1**, 100 (1861).
(103) Griess, P., *Ann.* **121**, 257 (1862).
(104) Griess, P., *Ann.* **137**, 39 (1866).
(105) Griess, P., Duisberg, C., *Chem. Ber.* **22**, 2459 (1889).
(106) Griess, P., *Proc. Roy. Soc. London* **9**, 594 (1859).
(107) Griess, P., *Trans. Roy. Soc. London* **A154**, 667 (1864).
(108) Heumann, K., *J. Soc. Chem. Ind.* **1890**, 1121.
(109) Heyman, B., *Angew. Chem.* **44**, 797 (1931).
(110) Hoechst Co., German Patent **32277** (Nov. 25, 1884).
(111) Hoechst Co., German Patent **40377** (Nov. 4, 1886).
(112) Hoechst Co., German Patent **38573** (March 12, 1887).
(113) Hofmann, A. W., *Ann.* **115**, 248 (1860).
(114) Hofmann, A. W., *Chem. Ber.* **10**, 388 (1877).
(115) Holderman, K., *Angew. Chem.* **45**, 141 (1932).
(116) Japp, F. R., *J. Chem. Soc.* **1898**, 97.
(117) Johnson, M., *J. Soc. Chem. Ind.* **40**, 176T (1921).
(118) Julius, P., *Angew. Chem.* **24**, 2417 (1911).
(119) Julius, P., *Angew. Chem.* **38**, 737 (1925).
(120) Julius, P., *Chem. Ber.* **19**, 1365 (1886).
(121) Julius, P., *Chem. Ber.* **19**, 2549 (1886).
(122) Kekulé, A., *Ann.* **137**, 129 (1866).
(123) Kekulé, A., *Ann.* **221**, 230 (1883).
(124) Kekulé, A., *Chem. Ber.* **2**, 748 (1869).
(125) Kekulé, A., Hidegh, C., *Chem. Ber.* **3**, 233 (1870).
(126) Kekulé, A., Franchimont, A., *Chem. Ber.* **5**, 906 (1872).
(127) Kekulé, A., *Compt. Rend.* **64**, 752 (1867).
(128) Kekulé, A., *Z. Chem.* **2**, 687 (1866).
(129) Kekulé, A., *Z. Chem.* **2**, 689 (1866).
(130) Kekulé, A., *Z. Chem.* **2**, 693 (1866).
(131) Kekulé, A., *Z. Chem.* **2**, 700 (1866).
(132) Kekulé, A., "Chemie der Benzol Derivate," Vol. 1, pp. 196-252, Ferdinand Enke, Erlangen, 1867.
(133) Kekulé, A., "Lehrbuch der Organischen Chemie," Vol. 2, pp. 688-744, Ferdinand Enke, Erlangen, 1866.
(134) Knorr, L., *Ann.* **238**, 137 (1887).
(135) Knorr, L., *Chem. Ber.* **17**, 5491 (1884).
(136) Knorr, L., German Patent **26429** (July 22, 1883).
(137) Körner, W., *Ann.* **137**, 197 (1866).
(138) Körner, W., *Gazz. Chim. Ital.* **4**, 305 (1874).
(139) Körner, W., *Giorn. Accad. Sci. Econ. Natur. Palermo* **5**, (April 20, 1869).
(140) Ladenburg, A., *Chem. Ber.* **9**, 219 (1876).
(141) Lieberman, C., Graebe, C., U.S. Patent **95465** (Oct. 5, 1869).
(142) Lieberman, C., Graebe, C., British Patent **3850** (Dec. 18, 1868).
(143) Luttringhaus, A., *Angew. Chem.* **44**, 109 (1931).
(144) Matthews, T. M., *J. Soc. Chem. Ind.* **20**, 551 (1901).
(145) Martius, C. A., Griess, P., *Z. Chem.* **2**, 132 (1866).
(146) Mohring, H., Bayer, Leverkusen private communication.
(147) Oehler, K. and Co., German Patent **47235** (April 25, 1888).
(148) Oehler, K. and Co., British Patent **7997** (May 31, 1888).
(149) Partington, J. R., "A History of Chemistry," Vol. 4, Macmillan, London, 1964.
(150) Pechmann, H., Duisberg, C., *Chem. Ber.* **16**, 2119 (1883).
(151) Perkin, W. H., *J. Soc. Chem. Ind.* **4**, 427 (1885).
(152) Perkin, W. H., *J. Soc. Chem. Ind.* **25**, 783 (1906).
(153) Perkin, W. H., *J. Chem. Soc.* **1923** 1520.

(154) Poggendorff, "Handworterbuch," Vol. 8, p. 698, Leipzig, 1863.
(155) Rassow, B., *Angew. Chem.* **44,** 879 (1931).
(156) Richter, V., *Chem. Ber.* **21,** 2476 (1888).
(157) Richter, V., "Organic Chemistry," transl. by E. F. Smith, Vol. 2, pp. 526-28, P. Blakiston's Son & Co., Philadelphia, 1900.
(158) Roberts, Dale, and Co., *J. Soc. Chem. Ind.* **4,** 476 (1885).
(159) Robiquet, P. J., Colin, J. J., *Ann. Chim.* **34,** 225 (1827).
(160) Robiquet, P. J., *J. Pharm. Alsace Lorraine* **21,** 387 (1835).
(161) Rowe, F. M., *J. Soc. Dyers Colourists* **51,** 218 (1935).
(162) Rudolph, C., Priebs, B., U.S. Patent **396,294** (Jan. 15, 1889).
(163) Runge, F. F., *Ann. Phys.* **31,** 75 (1834); **32,** 308 (1834).
(164) Sansone, A., *J. Soc. Chem. Ind.* **4,** 18 (1885).
(165) Schmitt, R., *Ann.* **112,** 118 (1859).
(166) Schofield, M., *Dyer,* **125,** 161 (1961).
(167) Schultz, G., *Ann.* **174,** 201 (1874).
(168) Schultz, G., *Chem. Ber.* **23,** 1265 (1890).
(169) Schultz, G., Julius, P., *Chem. Ind. (Dusseldorf)* **10,** 295 (1887).
(170) Schultz, G., Julius, P., *J. Soc. Chem Ind.* **6,** 653 (1887).
(171) Schultz, G., *J. Soc. Dyers Colourists* **44,** 181 (1928) [obituary].
(172) Schuster, C., "Badische Anilin- und Soda-Fabrik AG" (no date).
(173) Society of Dyers and Colourists and American Association of Textile Chemists and Colorists, "Colour Index," 2nd ed., The Society, Bradford, 1956.
(174) Stevenson, J. C., *J. Soc. Chem. Ind.* **4,** 438 (1885).
(175) Strecker, A., *Chem. Ber.* **4,** 784 (1871).
(176) Strecker, A., *Z. Chem.* **7,** 481 (1871).
(177) Suida, W., *Chem. Ber.* **11,** 584 (1878).
(178) Wagner, R., *Chem. Ber.* **5,** 125 (1872).
(179) Weinberg, A., *Angew. Chem.* **43,** 167 (1930).
(180) Wilcox, D. H., Jr., *Am. Dyestuff Rep.* **47,** 539 (1958).
(181) Wilcox, D. H., Jr., *Am. Dyestuff Rep.* **53,** 15 (1964).
(182) Wilcox, D. H., Jr., Greenbaum, F. R., *J. Chem. Ed.* **42,** 266 (1965).
(183) Wilcox, D. H., Jr., Greenbaum, F. R., "The Development of the Structure of Benzene or a Translation of the Five Pertinent Papers by August Kekule published in 1865, 1866, 1869, and 1872," in press.
(184) Witt, O. N., *Chem. Ber.* **9,** 522 (1876).
(185) Witt, O. N., *Chem. Ber.* **10,** 654 (1877).
(186) Witt, O. N., *J. Soc. Chem. Ind.* **34,** 409 (1915) [obituary].
(187) Wolff, J., Strecker, A., *Ann.* **75,** 1 (1850).
(188) Wurtz, A., *Compt. Rend.* **64,** 749 (1867).
(189) Ziegler, J. H., Loscher, M., *Chem. Ber.* **20,** 834 (1887).
(190) Ziegler, J. H., U.S. Patent **324,630** (Aug. 18, 1885).

RECEIVED September 24, 1965.

4

The Spatial Configuration of the Benzene Molecule and the End of the Kekulé Formula

A. SEMENTSOV

Lafayette College, Easton, Pa.

Three-dimensional models of benzene correspond to the structural formulas with double, diagonal, and centric bonds. The most interesting is the formula of Sachse, which corresponds to the formula containing three-electron bonds. The recently synthesized three-dimensional isomers of benzene are not aromatic, and the less stable, the more double bonds they contain. Hence, the Kekulé formula with three double bonds can not express the benzene properties and is only a "Bildungsformel." It is accepted that hydrogen atoms are located in the same plane which contains carbon atoms, but some experimental data contradict this presumption.

Kekulé's formula for benzene was criticized almost as soon as it was published. Alternative formulas were proposed, beginning in 1867 when Claus (5, 6, 7) advanced his well known diagonal formula. Some of these formulas were three dimensional.

Ladenburg (21) wrote in 1869 that he had told Kekulé several years before that the positions 1,2 in his formula are not identical with the positions 1,6. Markovnikov (27), in his thesis published in the same year, does justice to the enormous usefulness of the Kekulé formula. However, he suggests that it is much less important than Kekulé's hypothesis about the structure of saturated compounds. Furthermore, he assumes, as did Kekulé (19), that only the synthesis of benzene can prove this formula. It is interesting to note that all benzene systeses correspond to the Kekulé formula, but they did not confirm it. We will see this later.

The properties of benzene limit the configuration of its molecule to three possibilities—hexagon, triangular prism, and octahedron. Hence, all three-dimensional models must be reduced to the prism and octahedron.

Both of these models were suggested in 1869. Koerner (*20*) proposed the model shown in Figure 1.

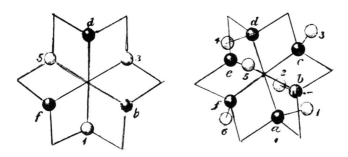

Figure 1.   Koerner's model

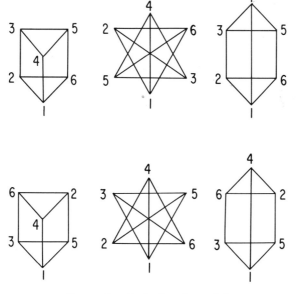

Figure 2.   Ladenburg's models

Ladenburg (*21*) proposed the prismatic model and also a model in the form of a twisted prism. The last one, as well as Koerner's model, was octahedral (Figure 2). The original numbering of Ladenburg was later changed to that in the lower left of Figure 2 to meet the experimental data.

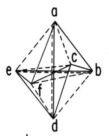

THOMSEN'S SYMBOL (continuous line)
MEYER'S SYMBOL (dotted line)

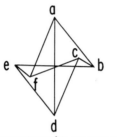
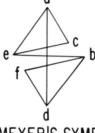

THOMSEN'S SYMBOL   MEYER'S SYMBOL

Figure 3.   Octahedral models of Thomsen and Meyer

Octahedral models were also offered by Richard Meyer (*29*) and Thomsen (*37*) (Figure 3). The only difference between them is the direction of the axial bonds. Meyer proceeded from the model of Ladenburg while Thomsen kept the direction of the diagonal bonds in Claus' formula.

Sworn (*36*) in 1889 strongly favored the octahedral model of Thomsen. He assumed that this configuration ensures the most stable equilibrium and explains the remarkable fact that only molecules containing six atoms show aromatic properties. He also emphasizes the necessity of diagonal bonds to give the molecule the needed compactness and explains the difference of the ortho and para positions from the meta ones. It is interesting to note that Pauling (*32*) used the same argument in favor of diagonal bonds as late as 1926. Sworn also proposed a projection which is not octahedral but prismatic (Figure 4).

In 1888 Herrmann (*15*) proposed an octahedral model with the positions of hydrogen atoms different from all others (Figure 5).

Much later, Collie (*8*) suggested a dynamic octahedral model, in which two octahedral configurations are transformed into each other through an intermediate flat hexagonal configuration. He repeated this idea again in 1916 (9) and tried to confirm his model by the study of the ultraviolet spectra of benzene (2).

All these models contradict the fact that carbon atoms in the benzene molecule are located in one plane, or almost in one plane. That was

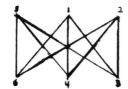

Figure 4.   Sworn's prismatic model

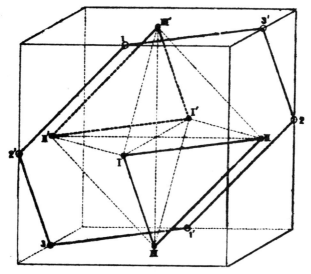

Figure 5.   Herrmann's octahedral model

established by x-ray diffraction (10, 26), electron diffraction (33, 41), infrared and Raman spectra (17). In all models mentioned, the authors consider the carbon atoms as material points.

Seven authors proposed models constructed of tetrahedral carbon atoms. In these models, we see the application of the van't Hoff theory to the formulas of Kekulé, Armstrong-Bayer, and those with the three-electron bonds (Figure 6).

The model corresponding to Kekulé's formula was considered and rejected by Marsh (28) and defended by Graebe (14). Marsh rejected it because it did not express the specificity of aromatic compounds—e.g., in the sharp change of properties by the dihydrogenation of phthalic acid.

Marsh, Loschmidt (25), Erlenmeyer (12), Vaubel (39), and Huggins (16) proposed the tetrahedral interpretation of the Armstrong-Bayer formula. In the first three models, the centers of six carbon tetrahedra are in the same plane. Baeyer (1) adhered to this model. In the models of Vaubel and Huggins, the centers of tetrahedra are located alternately in two parallel planes.

Graebe (14) criticized all these models, claiming that they cannot explain why only phthalic acid, but not iso- and terephthalic acids, gives an anhydride. He indicates that it is difficult, if not impossible, to construct models for condensed aromatic substances, proceeding from these models.

Another fact which contradicts the models of Vaubel and Huggins is the known planarity of the benzene nucleus. The most interesting is the model proposed by Sachse (34) in 1888. In this model every carbon atom tetrahedron is bound to two neighboring tetrahedra by an edge. It implies that each C—C bond contains three electrons.

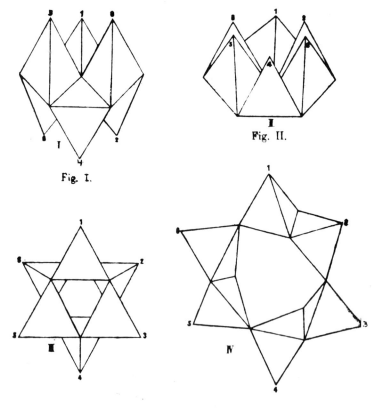

Figure 6.   Tetrahedral carbon models

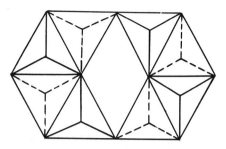

Figure 7.   Sachse's model

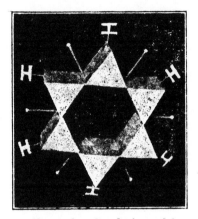

Figure 8.   Cracker's model

Formulas containing three-electron bonds were suggested by **Kauff-**man in 1911 and Thomson in 1914 and again in 1921.

Recently, Linnett (*24*) proposed a new benzene formula with three-electron bonds. It is based on the assumption that the electron octet consists of two tetrahedra, one of which contains electrons with a positive spin, the other with negative spin.

In the Sachse model, the electron positions correspond to the Linnett formula (Figure 7). It is interesting to note the way Huggins explains the lack of aromatic properties in cyclooctatetraene. Proceeding from Erlenmeyer's stereochemical interpretation of the conjugation, he assumes that in its molecule the cyclic conjugation is impossible.

Another model with three-electron bonds was proposed by Crocker (*11*) (Figure 8). Here, two neighboring carbon atoms are bound by the electron pair and an additional "aromatic" electron.

I will not describe the complicated model of Morse (*31*), based on the idea of the cubic form of the electron octet.

Several recently published papers describe polycyclic valence isomers of the benzene derivatives. All these isomers were nonaromatic and less stable than their aromatic isomers. The most interesting is that of Viehe and co-workers (*40*), who prepared all three possible polycyclic isomers (Figure 9).

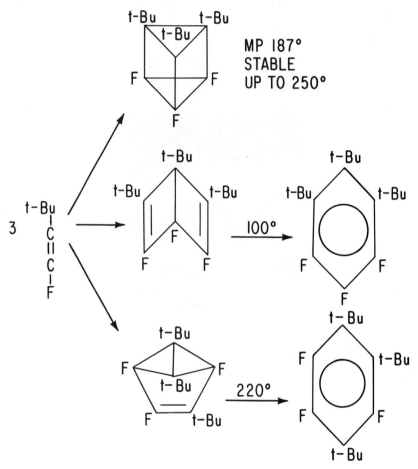

Figure 9. The three possible polycyclic isomers prepared by Viehe

The tetracyclic prismatic isomer which does not contain any double bonds is the most stable (up to 250° C.). The tricyclic isomer with only one double bond isomerizes into the aromatic isomer more easily at 220° C. The least stable was the so-called Dewar isomer with two double bonds, isomerizing at 100° C.

From this we can conclude: (1) all polycyclic (diagonal) models of benzene are wrong, and (2) the stability of isomers of benzene derivatives decreases with an increased number of double bonds. Therefore, the Kekulé formula, with its three double bonds, cannot express the benzene properties. However, the Kekulé formula corresponds perfectly to all syntheses of benzene. For this reason, Erlenmeyer (*12*) assumes that it is only a "bildungsformel." This suggests that Kekulé's benzene is formed by benzene syntheses, but it is unstable and is transformed, in *statu nascendi*, into a specific aromatic structure. In favor of this hypothesis is the radical change of properties by the transformation of dihydro derivatives of benzene into aromatic substances and vice versa. An example is the abrupt change of properties by introducing the third double bond by the Willstätter benzene synthesis. Perhaps the specific aromatic structure contains the three-electron bonds as Linnett proposes. It is interesting to note that Linnett's formula explains the composition of the mixture of products of ozonization of *o*-xylene and 1,2,4-trimethylbenzene.

In regard to the position of hydrogen atoms and their subsituents, only Graebe and Cracker assume that they are located in the plane of the carbon atoms ring. All the other models require that unsymmetrically substituted derivatives must be optically active. It is known that the natural aromatic substances which do not have asymmetry in side chains are not optically active. Several attempts to resolve unsymmetrically substituted benzenes into optical isomers failed (*22, 23*). The negative result of the attempt to resolve nitro- and formylthymotic acid (*30*) was particularly important because all six substituents in it are different. Hence, the plane of the ring is the only possible symmetry plane. For this reason, it is accepted that hydrogen atoms and substituents are located in the plane of the carbon atoms ring. However, some experiments cast doubt on this conclusion, at least in some cases.

The present author succeeded in resolving *o*-toluidine disulfonic acid (*35*). The possible cause of the difference in results, as compared with the results of Lewkovitsch and Meyer, is the difference in the method used. The enantiomeric salt was precipitated with an insufficient amount of optically active base.

The electron diffraction method (*4*) shows that in *o*-dichloro- and *o*-dibromobenzene the halogen atom lies out of the carbon atom plane. Ferguson and Slim (*13*) demonstrated by the x-ray method that the carboxyl groups in substituted benzoic acids are deflected up to 23° from the plane of the ring.

Literature Cited

(1) Baeyer, A., *Ann.* **245**, 123 (1888).
(2) Baly, E. C. C., Collie, J. N., *J. Chem. Soc.* **87**, 1332 (1905).
(3) Baly, E. C. C., Edwards, W. H., *J. Chem. Soc.* **89**, 514 (1906).
(4) Bastiansen, O., Hassel, O., *Acta Chem. Scand.* **1**, 489 (1947); *C. A.* **42**, 2484 (1947).

(5) Claus, A., "Theoretische Betrachtungen und deren Anwendung zur Systematic der Organischen Chemie," p. 207, Freiburg, 1867.
(6) Claus, A., *Ber.* **15**, 1907 (1882).
(7) Claus, A., *Ber.* **20**, 1423 (1887).
(8) Collie, J. N., *J. Chem. Soc.* **71**, 1013 (1897).
(9) Collie, J. N., *J. Chem. Soc.* **109**, 561 (1916).
(10) Cox, E. G., *Proc. Roy. Soc.* **A135**, 441 (1932).
(11) Crocker, E. C., *J. Am. Chem. Soc.* **44**, 1618 (1922).
(12) Erlenmeyer, E., *Ann.* **316**, 57 (1901).
(13) Ferguson, G., Slim, G. A., *Proc. Chem. Soc.* **1961**, 162.
(14) Graebe, C., *Ber.* **35**, 526 (1902).
(15) Herrmann, G., *Ber.* **21**, 1949 (1888).
(16) Huggins, M. L., *J. Am. Chem. Soc.* **44**, 1607 (1922).
(17) Ingold, C. K., *et al.*, *J. Chem. Soc.* **1936**, 912, 915, 925, 931, 941, 955, 966, 971; C. A. **30**, 7033 (1936).
(18) Kauffmann, H., "Die Valenzlehre," p. 535, F. Enke, Stuttgart, 1911.
(19) Kekulé, F. A., *Ann.* **162**, 77 (1872).
(20) Koerner, W., *Giorn. Sci. Nat. Ed. Econ.* V, 241 (1869); cited from *J. Chem. Soc.* **29**, 241 (1876).
(21) Ladenburg, A., *Ber.* **2**, 140, 272 (1869).
(22) LeBel, A., *Bull. Soc. Chem. France* **38**, 98 (1882).
(23) Lewkowitsch, J., *J. Chem. Soc.* **53**, 781 (1888).
(24) Linnett, J. W., *Am. Scientist* **52**, 470 (1964).
(25) Loschmidt, J., *Monatsh.* **11**, 28 (1890); *Chem. Zentr.* **1891**, I, 786.
(26) Mark, H., *Ber.* **57**, 1820 (1929).
(27) Markovmikov, B. B., "Izbrannye Trudy" (Selected Works), p. 234, ANSSSR Moskow, 1955.
(28) Marsh, E., *Phil. Mag.* **26**, 426 (1888).
(29) Meyer, Richard, *Ber.* **15**, 1823 (1882).
(30) Meyer, V., Luehr, F., *Ber.* **28**, 794 (1895).
(31) Morse, J. K., *Proc. Natl. Acad. Sci.* **13**, 789 (1927).
(32) Pauling, L., *J. Am. Chem. Soc.* **48**, 1132 (1926).
(33) Pauling, L., Brockway, L. O., *J. Chem. Phys.* **2**, 867 (1934).
(34) Sachse, H., *Ber.* **21**, 2530 (1888).
(35) Sementsov, A., *Ukr. Khim. Zh.* **8**, 193 (1934).
(36) Sworn, S. A., *Phil. Mag.* **28**, 402, 443 (1889).
(37) Thomsen, J., *Ber.* **19**, 2944 (1886).
(38) Thomson, J. I., *Phil. Mag.* **27**, 757 (1914).
(39) Vaubel, W., *J. Prakt. Chem.* **44**, 137 (1891).
(40) Viehe, H. C., Messingi, R., Ott, J. F. M., Senders, J. R., *Angew. Chem. Intern. Ed. Engl.* **3**, 695 (1864).
(41) Wierl, R., *Ann. Phys.* **8**, 521 (1931).

RECEIVED September 24, 1965.

The Development of the Understanding of Unsaturation: 1858–1870

A. ALBERT BAKER, JR.

Grand Valley State College, Allendale, Mich.

The central problem in the study of unsaturated substances resolved itself into the question: what unique feature do these compounds possess, which allows some to behave as they do? A satisfactory answer was found between 1858 and 1870, growing out of the concept of carbon tetravalency and graphic representation of how bonded atoms are arranged in molecules. Kekulé and Couper provided the former simultaneously in 1858, and at the same time Couper suggested a method for graphic representation. Couper's suggestion was developed by Crum Brown's explaining isomerism, adapted by Erlenmeyer, and made generally applicable by Butlerov, who demonstrated that the assumption of multiple bonds was not only compatible with chemical behavior of unsaturated compounds but necessary to explain that behavior.

The unsaturated hydrocarbon, ethylene, was first prepared by the Dutch chemists in 1794 (*16*). By 1850 at least 12 hydrocarbons exhibiting properties similar to those of ethylene were known, and during the following decade many new unsaturated substances were identified.

The central problem in studying unsaturated substances was finding what unique feature these compounds possess which allows them to behave as they do? A satisfactory answer to this problem arose from the concepts of the tetravalency of the carbon atom and graphic representation of how atoms in a molecule are arranged through their bonding.

Valency of Carbon

The concept of the tetravalency of carbon was provided simultaneously by August Kekulé (*22*) and Archibald Scott Couper (*15*) in 1858. Couper also suggested a method of depicting graphically the arrangement of atoms within a molecule, but Charles Wurtz (24) objected to Couper's graphic formulas, saying they were too arbitrary and were too far removed from experience whereas the rational formulas of Gerhardt had the advantage of representing only the metamorphoses of chemical compounds without becoming involved in the hypothetical positioning of each atom within the molecules. This objection, however, did not prevent Wurtz from devising his own method of graphic representation six years later (25), when he proposed that ethylene was composed of a divalent carbon atom united with a tetravalent carbon atom, having the formula $CHCH_3$ (Figure 1).

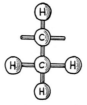

Figure 1.
Wurtz's formula
for ethylene
(top) and
Hofmann's
"unfinished"
molecule
(bottom)

The idea of divalent carbon, which had been suggested as a possibility by Couper in 1858 and which was not at first ruled out by Kekulé, proved to be a stumbling block in understanding unsaturation. August Wilhelm von Hofmann (21) perpetuated the concept of divalent carbon in his models of "unfinished molecules," and only after the termination of Alexander Butlerov's (3) long and unsuccessful search for methylene (CH_2), was the idea largely discarded.

The Concept of Chemical Structure

In 1859 a translation of Couper's paper, as it appeared in *Comptes rendus*, was published in *Annalen der Chemie* (13, 14), followed by Butlerov's remarks on Couper's new chemical theory (4). Butlerov was rather cautious about adopting a new symbolism for expressing chemical constitution, and Kekulé's ideas, which were framed within the Gerhardt type theory, were more to his taste. However, within two years Butlerov had devised the term "chemical structure" (chemische Structur) (5). In explaining the term, he proceeded from the basic assumption that each chemical atom bore a definite, limited amount of chemical force, or affinity, with which it participated in forming a molecule. He referred to the mutual uniting of atoms in a complex body as "chemical binding" and to the resulting complex body as "chemical structure."

In defining chemical structure, Butlerov tied together the concept of atoms, valence, and interatomic bonding. He was confident that once the correlations between chemical properties and chemical structures became known, such structures would express all the properties, but until such correlations were known, he felt unable to propose or accept a method for depicting chemical structure.

Isomerism and Structural Formulas

Alexander Crum Brown (1) proposed a method for depicting chemical structures in a paper read before the Royal Society of Edinburgh May 2, 1864 (2). His paper contained the first attempt to explain isomerism in organic compounds by using structural formulas that are clearly recognizable as the same type of structural formulas used today. At this time, isomerism was in an almost hopelessly confused state, complicated by the failure to recognize the identity of compounds obtained from different sources. For example, "methyl gas" (C_2H_6), obtained by treating methyl iodide with zinc, was believed to be different from "hydride of ethyl" (C_2H_6), obtained by treating ethyl iodide with zinc and water. Butlerov (6) tried to explain this difference by assuming that carbon could possess nonidentical valences—i.e., a primary valence and a secondary valence. He explained that primary valences were used in normal carbon-to-carbon linkages, but a carbon-to-iodine linkage in methyl iodide involved the secondary valence of the carbon atom. Thus, the carbon-to-carbon linkage in methyl gas would be composed of secondary valences since a new carbon-to-carbon linkage had been formed from the carbon-to-iodine linkages; however, the carbon-to-carbon linkage in hydride of ethyl would be composed of primary valences since the carbon-to-carbon linkage had already existed in ethyl iodide. Therefore, the two compounds having the same formula, would be isomeric but not identical.

Since he was unable to write more than one structural formula for C_2H_6, Crum Brown was forced to accept, reluctantly, Butlerov's views. Nevertheless, he advised a thorough investigation of the problem because if taken too far, the concept of nonidentical valences could lead to absurd results. A few months later Carl Schorlemmer (23) of Manchester demonstrated that methyl gas and hydride of ethyl were clearly the same compound—ethane. His work prompted chemists to reinvestigate the properties of many compounds which had previously been believed to be isomers, and many pairs of so-called isomers were found to be one and the same compound.

By using structural formulas, Crum Brown was able to demonstrate the difference between propyl alcohol proper (primary propyl alcohol) and Friedel's alcohol (secondary propyl alcohol) (Figure 2). He assigned them structures on the basis of the ability of the former to form

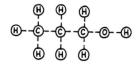

Figure 2.   Crum Brown's propyl alcohol (top), Friedel's alcohol (middle), and vinyl chloride (bottom)

an aldehyde and the latter to form a ketone. He also wrote structures for the two propyl iodides: "iodode of propyl" (primary propyl iodide) and "hydriodate of propylene" (secondary propyl iodide). He obtained the former by treating propyl alcohol with phosphorus iodide; the latter was prepared by adding hydrogen iodide to propylene, and the subsequent reaction with potassium hydroxide produced Friedel's alcohol. Having established the structures of propyl alcohol and Friedel's alcohol, which he also called hydrate of propylene, Crum Brown said that it was highly probable that all alcohols obtained from olefins would be similar to Friedel's alcohol—i.e., secondary alcohols.

In taking up the problem of the isomeric compounds, vinyl chloride and chloracetene (both possessing the formula C_2H_3Cl), Crum Brown was able to draw only one structural formula, and the formula he wrote was very likely the first representation of a chemical double bond (Figure 2). Couper had suggested multiple bonds, and Kekulé and Wurtz had indicated them with their circles and boxes, but with all three the primary object had been arithmetic; they needed to use up each carbon atom's four valence units. Crum Brown's approach was chemical, and although he didn't fulfill Butlerov's hopes that chemical structures would indicate all chemical properties, he was moving in that direction. In proposing his structural formula for vinyl chloride (chloracetene was later found to be nothing more than a mixture of phosgene, acetaldehyde, and paraldehyde), Crum Brown said that he would not deny the possibility of divalent carbon. However, all that was known of olefinic compounds led him to believe that the valence of the carbon was reduced, not by one or more of the carbon atoms' becoming divalent, but by the union of two carbon atoms sharing two valence units each rather than one. Thus, Crum Brown proposed, in effect, that the unique feature of unsaturated compounds is the sharing of two valence units by each of two carbon atoms, symbolized by the double bond.

Ethylene Compounds

Crum Brown strengthened his argument by demonstrating consistencies between his formulas for ethylene, ethylene dichloride, and ethylidene chloride and their chemical properties. Ethylene dichloride (1,2-dichloroethane), produced by chlorinating ethylene, could be converted into ethylene oxide whereas ethylidene chloride (1,1-dichloroethane) could be converted, not into an oxide, but into acetaldehyde. If one agreed with the generally held view that the oxygen atom in acetaldehyde was attached to only one carbon atom, he could scarcely find fault with Crum Brown's reasoning.

Chemists in general were very hesitant in accepting and using graphic representations of chemical constitution. Skepticism about the real existence of physical atoms and the uncertainty of the correlation between properties and structure were contributing factors. However, there were a few men who made immediate and fruitful use of structural formulas patterned after those used by Crum Brown. Among them, Edward Frankland and Baldwin Francis Duppa (*19*) at the Royal Institution were able to differentiate between the isomers of their newly synthesized hydroxy acids. Two other Englishmen, Ernest Theophron Chapman and William Thorp (*12*), enthusiastically adopted Crum

Brown's formulas, but they failed to advance the understanding of unsaturation when they proposed that olefins consisting of five or more carbon atoms were substituted cyclopropanes rather than open chains containing double bonds.

Frankland

Frankland presented a convincing case for Crum Brown's ethylene structure in his interpretation of his syntheses of lactic and paralactic acids (*see* Figure 3). The former was prepared from the hydrolysis of ethylidene cyanohydrin and the latter from the hydrolysis of ethylene cyanohydrin. His line of reasoning was (1) acetaldehyde reacts with

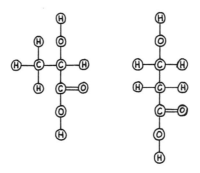

Figure 3. Frankland's formula for lactic acid (left)
and paralactic acid (right)

phosphorus pentachloride to form ethylidene chloride (1,1-dichloroethane); (2) ethylidene chloride hydrolyzes to form acetaldehyde; (3) since acetaldehyde and ethylene oxide are the only known compounds having the formula C_2H_4O, it can safely be assumed that in the former, oxygen is doubly bound to one carbon atom because the aldehyde can be prepared from alcohol by gentle oxidation whereas the oxide cannot be prepared that way; (4) if a double bond exists between the two carbon atoms in ethylene, it can react with cyanic acid to produce ethylene cyanohydrin but not ethylidene cyanohydrin (analogous to the production of ethylene dichloride by chlorinating ethylene); (5) the hydrolysis of ethylene cyanohydrin produces paralactic acid whereas the hydrolysis of ethylidene cyanohydrin produces lactic acid (*see* Figure 4). Frankland looked upon the conversion of acetaldehyde into ethylidene chloride as the simple replacement of divalent oxygen by two atoms of monovalent chlorine. With the structure of the ethylidene group thus demonstrated, he said that the only possible formula for ethylene was that in which the two carbon atoms were doubly bound.

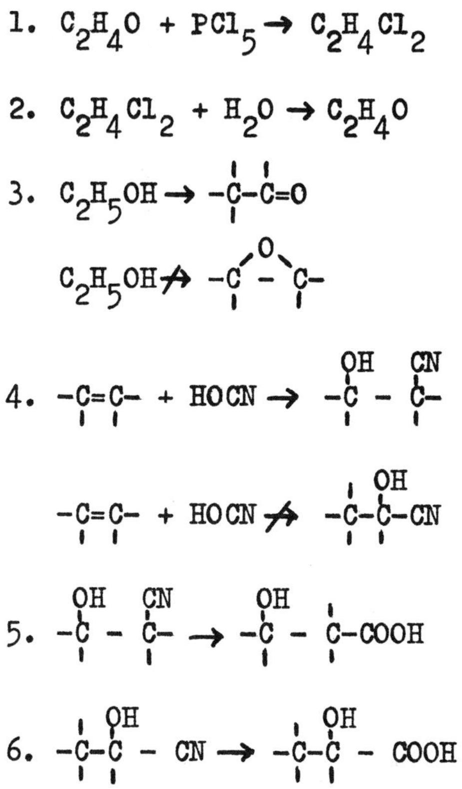

1. $C_2H_4O + PCl_5 \rightarrow C_2H_4Cl_2$

2. $C_2H_4Cl_2 + H_2O \rightarrow C_2H_4O$

3. $C_2H_5OH \rightarrow -\overset{|}{\underset{|}{C}}-\overset{|}{C}=O$

   $C_2H_5OH \nrightarrow -\overset{|}{C}\overset{O}{\diagdown}\overset{}{\underset{|}{C}}-$

4. $-\overset{|}{\underset{|}{C}}=\overset{|}{\underset{|}{C}}- + HOCN \rightarrow -\overset{OH}{\underset{|}{\overset{|}{C}}} - \overset{CN}{\underset{|}{\overset{|}{C}}}-$

   $-\overset{|}{\underset{|}{C}}=\overset{|}{\underset{|}{C}}- + HOCN \nrightarrow -\overset{|}{\underset{|}{C}}-\overset{OH}{\underset{|}{\overset{|}{C}}}-CN$

5. $-\overset{OH}{\underset{|}{\overset{|}{C}}} - \overset{CN}{\underset{|}{\overset{|}{C}}}- \rightarrow -\overset{OH}{\underset{|}{\overset{|}{C}}} - \overset{|}{\underset{|}{C}}-COOH$

6. $-\overset{|}{\underset{|}{C}}-\overset{OH}{\underset{|}{\overset{|}{C}}} - CN \rightarrow -\overset{|}{\underset{|}{C}}-\overset{OH}{\underset{|}{\overset{|}{C}}} - COOH$

Figure 4. Frankland's elucidation of structure in present-day notation

Erlenmeyer's Formulas

By 1866 Emil Erlenmeyer, at Heidelberg, had discovered Crum Brown. After a long, critical look at Kolbe's modification of type formulas, Erlenmeyer (17) found them lacking and ceased using them altogether. He used molecular formulas and something resembling present day line notation formulas. He recognized the success of Crum Brown's structural formulas, and he adapted them to his own use, omitting the circles around the symbols for atoms (18) (see Figure 5). Erlenmeyer consistently used the double bond in olefinic compounds, and he introduced the use of the triple bond in acetylene. He immediately adopted Kekulé's recently published proposal for the cyclic structure of benzene and also drew a structural formula for the unknown "diacetylene" (cyclobutadiene), a compound not synthesized for another 99 years (1965).

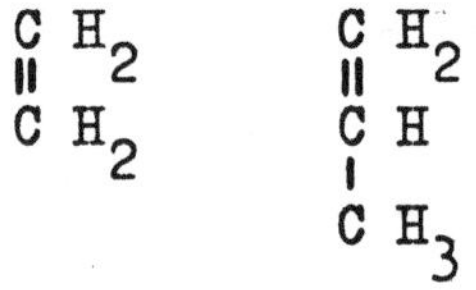

$$
\begin{array}{ll}
\text{C} & \text{H}_2 \\
\text{||} & \\
\text{C} & \text{H}_2
\end{array}
\qquad
\begin{array}{ll}
\text{C} & \text{H}_2 \\
\text{||} & \\
\text{C} & \text{H} \\
\text{|} & \\
\text{C} & \text{H}_3
\end{array}
$$

*Figure 5.   Erlenmeyer's formulas for ethylene (left)
and propylene (right)*

Superficially, Erlenmeyer's structural formulas looked more like those of Couper than those of Crum Brown. He did not draw in the lines representing bonds between carbon atoms and hydrogen atoms, assuming the reader knew they were intended. For example, his formula for ethane was $H_3C—CH_3$. In 1868, the year after Kekulé left the University of Ghent to go to Bonn, his former colleagues at Ghent, Professors Glaser and Swarts (20), adopted Erlenmeyer's formulas. Thus Couper's formulas, having passed through the mutations brought about by Crum Brown and Erlenmeyer, arrived at the birthplace of his rival's chemical theories after a lapse of 10 years.

Butlerov and Unsaturation

Still Butlerov hesitated. Scotch and English chemists were using structural formulas; Erlenmeyer and the professors at Ghent were using them; Kekulé and Wurtz had flirted with the idea but then backed away, and Butlerov waited. When he first coined the term "chemical structure" in 1861, Butlerov felt there was not yet enough solid evidence of the relationship between chemical properties and the positions of atoms within a molecule to permit the accurate portrayal of molecular structure. As time passed, he became thoroughly familiar with the problems connected with suitable structures for both the ethylenic and acetylenic types of unsaturation. These problems, coupled with those presented by isomerism, were leading him to a method of expressing structure which would, as far as possible, embody the qualities he felt structural formulas should possess.

Butlerov's work on *tert*-butyl alcohol led to his preparation of what he believed to be isobutylene, $CH_2C(CH_3)_2$ (7), and he was able to confirm this in 1869 (8). He stated the assumption that in an unsaturated compound of the type C_nH_{2n} there were doubly bound carbon atoms; therefore from isobutyl alcohol, only isobutylene could be formed, and only from such a form of butylene could *tert*-butyl alcohol be formed by adding water. He went on to show that the primary alcohol (isobutyl alcohol) could be converted into *tert*-butyl alcohol by first forming the iodide by means of concentrated hydriodic acid (9). The isobutyl iodide was then treated with alcoholic potassium hydroxide to produce isobutylene, which added hydrogen iodide to form *tert*-butyl iodide. Hydrolysis of the tertiary iodide resulted in the formation of *tert*-butyl alcohol. Thus, Butlerov saw the need to assume the presence of a carbon-to-carbon double bond in unsaturated hydrocarbons of the C_nH_{2n} series. No other assumption could explain the foregoing reactions. Such an assumption also explained why methylene could not exist. Unsaturation requires a double bond between two carbon atoms; therefore, a compound containing only one carbon atom per molecule cannot be unsaturated.

In 1870 Butlerov took the final step of incorporating the double bond into structural formulas (*10*, Figure 6):

Amongst the different hypotheses brought forward to explain the constitution of these bodies, the author prefers that which assumes that two atoms of carbon are linked together by more than one unit of their combining capacity. This theory explains fully all the well known cases of isomerism in the olefine series; thus we know only one ethylene, one propylene, but three butylenes.

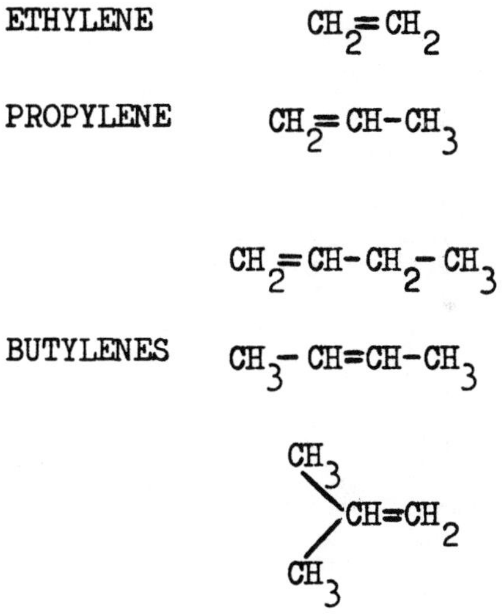

ETHYLENE $\qquad$ $CH_2 = CH_2$

PROPYLENE $\qquad$ $CH_2 = CH-CH_3$

$CH_2 = CH-CH_2-CH_3$

BUTYLENES $\qquad$ $CH_3-CH=CH-CH_3$

$$\begin{matrix} CH_3 \\ \diagdown \\ \diagup \quad CH=CH_2 \\ CH_3 \end{matrix}$$

Figure 6.   Butlerov's formulas for ethylene, propylene, and butylenes

Butlerov believed that the correctness of this theory could be proved by experiment because there should be alcohols and halides which would not yield olefins by abstracting water or hydrogen halide. As an example, he took neopentyl alcohol (2,2-dimethyl-1-propanol) which on dehydration could not yield a double bonded compound but could only produce a compound containing a three-membered ring. Butlerov (*11*) also felt that if an unsaturated compound could exist without double bonds, he should be able to prepare an isomeric propylene of the type $(CH_2)_3$, but all attempts to produce such a compound failed. It did not seem to occur to Butlerov that he might produce cyclopropane, although he was aware that a substituted cyclopropane could exist since he had stated that he obtained one from the dehydration of neopentyl alcohol.

With the publication of Butlerov's paper on the double bond as an expression of the property of unsaturation, chemists in ever increasing numbers began using structural formulas, incorporating double bonds for the ethylenic hydrocarbons and triple bonds for acetylenic compounds. Among them were Dewar, Fittig, Schiff, and Tollens as well as Erlenmeyer and Kekulé. Thus the idea, which had first been suggested as a possibility by Couper in 1858, which was developed to explain certain types of isomerism by Crum Brown, and which was adapted by Erlenmeyer in 1866, was made generally applicable by Butlerov in 1870 by his demonstrations that the assumption of multiple bonds was not only compatible with the chemical behavior of unsaturated compounds but necessary to explain that behavior.

Literature Cited

(1) Brown, A. C., *Trans. Roy. Soc. Edinburgh* **23**, 707 (1864).
(2) Brown, A. C., Ph.D. Thesis, University of Edinburgh, 1861.
(3) Butlerov, A. M., *Bull. Soc. Chim. France* **1861**, 84.
(4) Butlerov, A. M., *Ann.* **110**, 51 (1859).
(5) Butlerov, A. M., *Z. Chem.* **4**, 549 (1861).
(6) Butlerov, A. M., *Z. Chem.* **5**, 301 (1862).
(7) Butlerov, A. M., *Bull. Soc. Chim. France* **5**, 30 (1866).
(8) Butlerov, A. M., *Z. Chem.* **6**, 236 (1870).
(9) *Ibid.*, pp. 237-239.
(10) Butlerov, A. M., *J. Chem. Soc.* **24**, 214 (1871).
(11) Butlerov, A. M., *Chem. Zentr.* **42**, 94 (1871).
(12) Chapman, E. T., Thorp, W., *J. Chem. Soc.* **19**, 477 (1866).
(13) Couper, A. S., *Compt. Rend.* **46**, 1157 (1858).
(14) Couper, A. S., *Ann.* **110**, 46 (1859).
(15) Couper, A. S., *Phil. Mag.* **16**, 114 (1858).
(16) Deiman, J. R., van Troostwyck, A. P., Bondt, N., Lauwerenburgh, A., *J. Phys.* **45**, 178 (1794).
(17) Erlenmeyer, E., *Z. Chem.* **6**, 728 (1863).
(18) Erlenmeyer, E., *Ann.* **137**, 327 (1866).
(19) Frankland, E., Duppa, B. F., *Trans. Roy. Soc. London* **156**, 309 (1866).
(20) Glaser, C., *Z. Chem.* **4**, 338 (1868).
(21) Hofmann, A. W., *Proc. Roy. Inst. Gr. Brit.* **4**, 401 (1866).
(22) Kekulé, F. A., *Ann.* **106**, 129 (1858).
(23) Schorlemmer, C., *J. Chem. Soc.* **2**, 262 (1864).
(24) Wurtz, C. A., *Rep. Chim. Pure* **1**, 52 (1859).
(25) Wurtz, C. A., "Lecons de Philosophie Chimique," p. 137, Paris, 1864.

RECEIVED December 14, 1965.

Nonclassical Aromatic Compounds

NORMAN C. ROSE[1]

Texas A&M University, College Station, Tex.

The molecular orbital calculations of E. Hückel on the relative electronic stability of different monocyclic, conjugated systems has led to the preparation of a number of new, non-benzenoid aromatic compounds. The preparation and properties are given for the more intensely studied of these monocyclic aromatic systems that obey Hückel's rule. A few of the related compounds that do not obey Hückel's rule and that do not display aromatic properties are discussed.

Benzene was first isolated by Michael Faraday in 1825 from the pyrolysis products of a fish oil (28). For many years after its discovery, chemists wrestled with the problem of drawing for benzene a structural formula whose features are in keeping with the properties of benzene. In 1865 Kekulé deduced that benzene had a six-membered cyclic structure (41, 42). In 1872, he suggested that an oscillation occurs between the two arrangements, Formulas I and I', in such a way that the bonds to a given carbon from the two adjacent carbons do not differ (43).

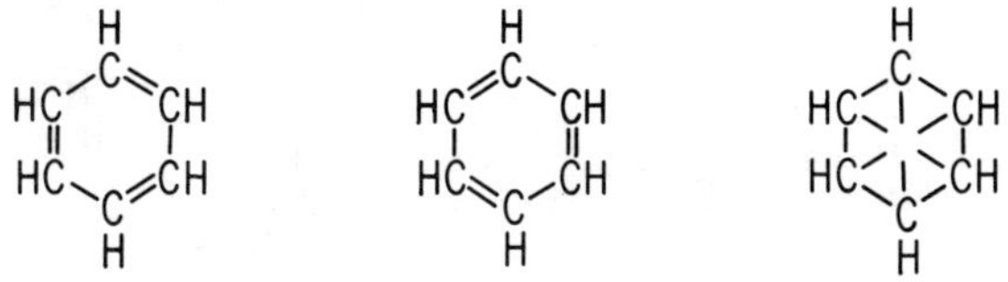

Although the formulations of Kekulé accounted for the cyclic nature of benzene, it is not apparent from these representations why benzene does not have the properties of an alkene. Claus and other investigators proposed special bonding schemes which do not have alkene-type linkages

---

[1] Present address: Portland State College, Portland, Ore.

(*17, 18*). These proposals were culminated by the Armstrong-Baeyer centric formula, II (*2, 3*). Neither the Claus nor the Armstrong-Baeyer formula readily accounts for the stability of benzene, nor can either of these types of formulas be extended easily to polynuclear aromatic compounds.

Bamberger used the centric formula, II, as a basis to correlate the aromatic character of pyrrole, and many other five-membered heterocyclic systems, with that of benzene (*4, 5*). He suggested that aromatic character was related to the presence of the six centric valences of the ring atoms. The four carbons of pyrrole each have one centric valence, and the nitrogen of pyrrole has two. According to Bamberger, the two salt-forming valences of the nitrogen of pyrrole must be incorporated into the centric group if pyrrole is to have the hexacentric character necessary for aromatic character. Bamberger postulated that the incorporation of the salt-forming valences of the nitrogen into the centric group resulted in pyrrole's being neutral. In contrast, pyrrolidine, which has no centric group, is basic.

Thiele in 1899 suggested that the stability of benzene could be accounted for by using his concept of partial valence (*58*). For a conjugated system, he used a curved line to represent the mutual neutralization of the partial valences on the carbons joined by a single bond. For example, the partial valences of butadiene would be indicated by:

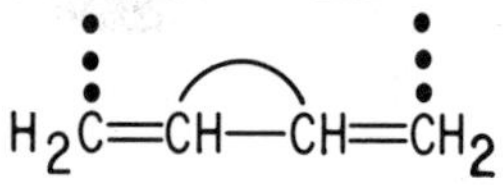

$$H_2C=CH-CH=CH_2$$

The dotted line is the partial valence of the terminal carbon. Benzene with its cyclic, conjugated system would be represented by III.

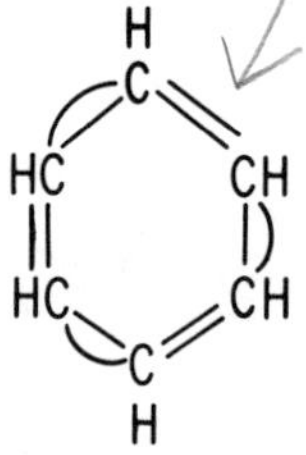

III 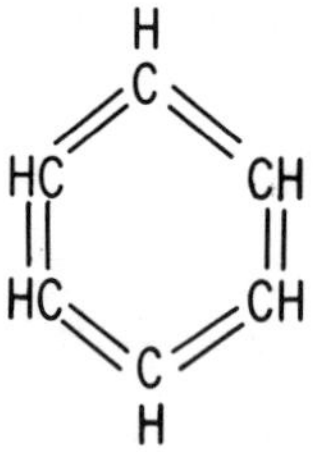   IV

Thiele said that since, through equalization of the partial valences, the original three double bonds have become inactive, one cannot distinguish between the three original double bonds and the three secondary double bonds; thus, benzene has six inactive double bonds. One could thus represent benzene by Formula IV.

In the early 1920's R. Robinson began to investigate "different" aromatic systems. He noted that six electrons in a cyclic, conjugated system form a group that resists disruption and that may be called the aromatic sextet (1). The concept of the aromatic sextet was used by other workers to explain the unusual acidity of cyclopentadiene since the anion formed would have an aromatic sextet (35).

In the 1930's the concept of resonance was developed and applied to molecules such as benzene. Explanations for the stability of benzene relative to 1,3,5-cyclohexatriene were based upon the phenomenon of resonance. However, this phenomenon could not be used to explain why aromatic stability depended upon six electrons per ring rather than on four or eight. An explanation of the stability of certain cyclic conjugated systems was developed by E. Hückel using simple molecular orbital theory (37). He calculated the π-orbital energies for monocyclic systems with various numbers of π-electrons. He found that the relative stability of conjugated, monocyclic systems depends upon the number of π-electrons in the system. His calculations have now been summarized by the expression $4n + 2$, usually referred to as Hückel's rule. The rule states that *conjugated, monocyclic* coplanar systems of trigonally hybridized atoms with $4n + 2$ π-electrons will possess relative electronic stability. These systems will have 2,6,10,14,18, etc. π-electrons—i.e., n equals 0,1,2,3,4, etc., respectively. Benzene, pyrrole, and furan are well known aromatic systems that have a number of π-electrons (six) in keeping with Hückel's rule.

In the closing years of the century following Kekulé's proposals there has been a revival of interest in aromatic compounds, particularly nonclassical aromatic systems. This renewed interest was generated by predictions of unusual stability for the cyclic systems implied in Hückel's rule.

Frost and Musulin suggested the following graphical device for representing the Hückel molecular orbital energies for the π-electrons for the cyclic systems on which Hückel based his calculations and for allied cyclic systems (32). A regular polygon of the ring is made with one of the ring atoms at the lowest point. Each atom of the ring is then projected horizontally. This automatically constructs the energy level diagram for the given ring and gives the correct energy scale if the radius of the ring is taken as 2β [β is the resonance integral (56)]. A horizontal line through the center of the ring corresponds to the zero resonance energy level. The vertical distance of each apex from the horizontal mid-line represents an energy level in units of β. Application of the mnemonic of Frost and Musulin to butadiene, benzene, and cyclooctatetraene is shown.

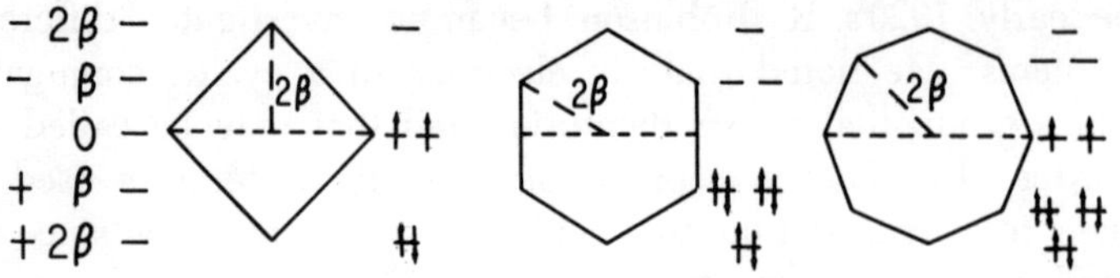

$$
\begin{array}{c}
-2\beta - \\
-\ \beta - \\
0 - \\
+\ \beta - \\
+2\beta -
\end{array}
$$

BUTADIENE   BENZENE   CYCLOOCTATETRAENE

Benzene has four sets of energy levels, of which two are filled. Each set may be thought of as a shell. The lowest and highest lying shells can accommodate two electrons while the remaining shells can accommodate four electrons. (The highest lying shell of some systems, for example, cyclopentadienyl anion (Formula V), is degenerate and could accommodate four electrons.) Thus, it takes $4n + 2$ electrons to yield *filled*-shell configurations.

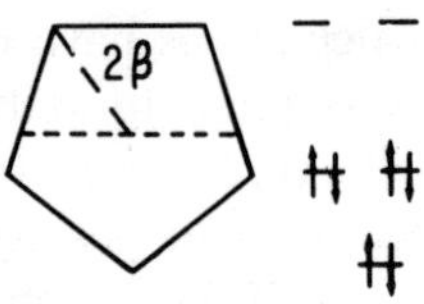

V

Simple molecular orbital calculations indicate that the delocalization energy of cyclobutadiene is 0, of benzene is 2β, and of cyclooctatetraene is 1.657β. For carbon systems, β is usually taken as 18 kcal./mole. The lack of aromatic properties for cyclooctatetraene indicates that the state of occupancy of the shells seems to have more importance than a net delocalization energy. Both cyclobutadiene and cyclooctatetraene have a half-filled shell of nonbonding molecular orbitals (orbitals of the zero resonance energy level).

Hückel's calculations were for monocyclic, conjugated systems. Attempts to apply Hückel's rule to polycyclic systems will not necessarily be successful. For example, pyrene (VI) is stable but has 16 π-electrons.

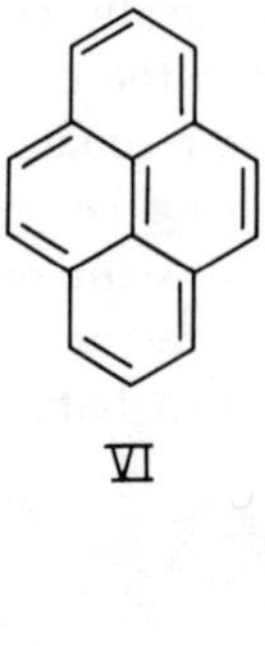

VI

The particular valence bond formulation given for pyrene can be viewed as a cyclic polyene having 14 electrons in the periphery and a cross-link containing π-electrons. Dewar suggested that such a polycyclic compound can be considered as a cyclic polyene having a small perturbation owing to the cross-link (*24, 25*). The $4n + 2$ rule can then be applied to the peripheral π-electrons. However, the cross-links are not small perturbations and predictions of possible aromatic character should not be expected to be valid in all cases.

Craig has proposed a rule to distinguish between aromatic and nonaromatic polycyclic systems (*19*). The rule is empirical since it is not based on energy calculations and uses the valence bond symmetry of the molecule as an index to whether delocalization does lead to stability. The rule can only be applied to hydrocarbons that have a symmetry axis passing through at least two or more π-centers. Rotation of the formula $180°$ about the symmetry axis must convert the original Kekulé form into itself or into another of the same canonical set. In order to apply the rule, each of the doubly bonded carbons of one of the Kekulé forms is labeled with the spin symbol α or β in such a way that the ends of each double bond have different spin symbols, and as few as possible like spin symbols are located on adjacent carbons. The sum of p and q is then determined. The symbol p is the number of interchanges of π-electron centers effected by the rotation, and q is the number of π-electron centers whose labels must be changed after the rotation in order to restore the original labeling scheme. If the sum $p + q$ is even, the valence bond ground state is symmetric, and the compound should be aromatic. If the sum is odd, the valence bond ground state is nontotally symmetric, and the compound should not be aromatic—i.e., the compound should be pseudoaromatic. Applications of Craig's rule are shown with VII, VIII, and IX.

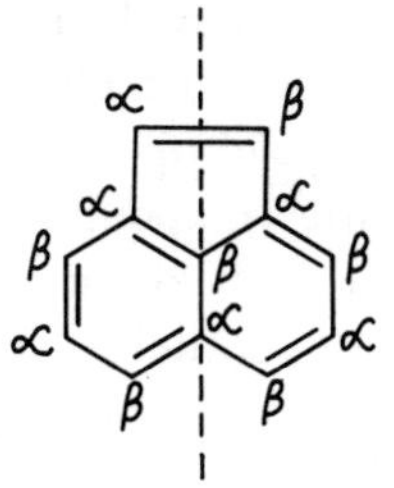

VII	VIII	I X
$p = 5$ $q = 1$	$p = 6$ $q = 0$	$p = 5$ $q = 0$
$p + q = 6$	$p + q = 6$	$p + q = 5$
Aromatic	Aromatic	Pseudoaromatic

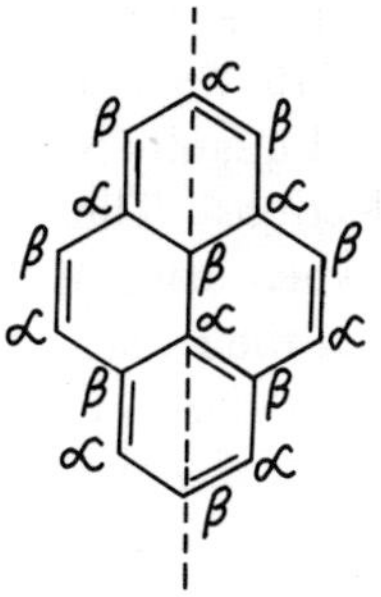

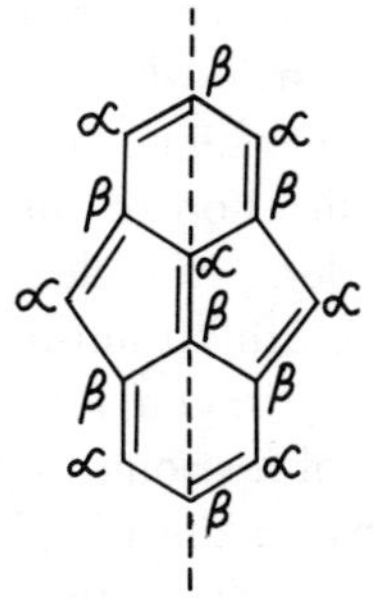

Craig believes Hückel's rule to be of limited applicability since ring systems obeying Hückel's rule must have at least a three-fold axis of symmetry (20). This limitation excludes a large majority of the conjugated systems. Craig's rule does not have this limitation. However, some ambiguity may arise in applying Craig's rule. The dimethyl derivative of aceheptylene (X) has aromatic properties (36). Applying Craig's rule to XI, one Kekulé form of the parent hydrocarbon, suggests it is aromatic; however, XII, a second Kekulé form of the parent hydrocarbon, appears to be pseudoaromatic by Craig's rule.

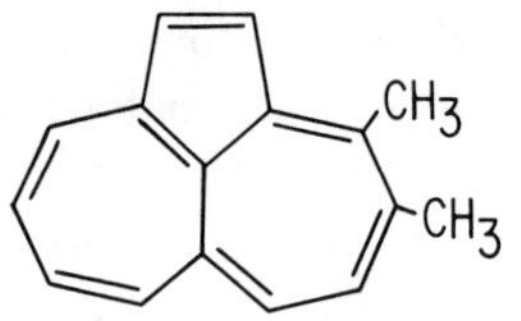

X

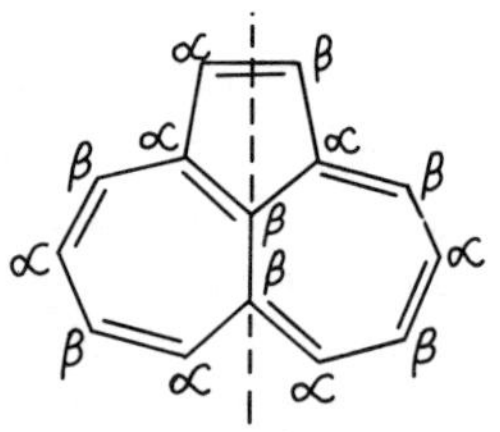

XI

$p = 6 \quad q = 1$
Pseudoaromatic

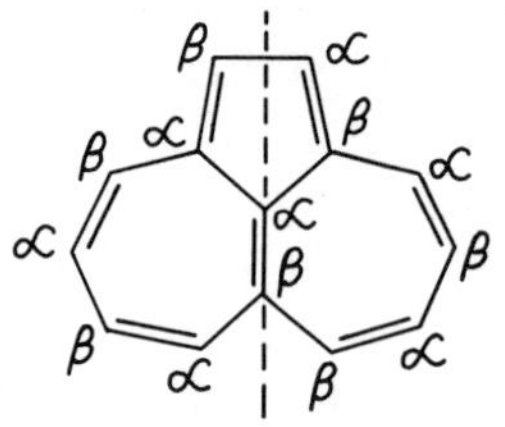

XII

$p = 6 \quad q = 6$
Aromatic

It is necessary to have some experimental criteria for aromaticity. In terms of classical concepts, chemical activity was often used as the criterion of aromatic character—that is, a compound was considered to be aromatic if it were unusually stable and underwent ionic electrophilic substitution reactions rather than ionic addition reactions. In the light of presently used concepts a monocyclic compound is considered to be aromatic if the compound has a cyclic, delocalized orbital that has a number of π-electrons in keeping with Hückel's rule, is reasonably planar, and has greater stabilization by π-electron delocalization than does an analog with localized bonds (16, 38). Presently, aromatic character is usually determined by physical quantities which depend upon the extent of delocalization of the π-electrons. Thus, aromatic compounds absorb light at long wavelengths when compared with alkenes, are readily polarized, exhibit anisotropy of their diamagnetic susceptibility and have a NMR hydrogen resonance in the low field region, indicative of the presence of a strong ring current of π-electrons.

Many of the nonclassical aromatic compounds are ions and do not lend themselves to the usual methods of quantitatively measuring resonance energies nor to electrophilic substitution reactions. Inferences that the compounds are aromatic have been made on the basis of their physical properties, particularly their NMR spectra. However, there are examples of nonclassical aromatic compounds which do undergo electrophilic substitution. An example will be presented later of the nitration of an oxabicycloundecapentaene, XXX.

$4n + 2 = 2$ Systems

Roberts, Streitwieser, and Regan determined the theoretically expected stability of a number of aromatic systems including the simplest aromatic system, the cyclopropenylium cation, XIII (*50*). The cyclopropenylium ion has two π-electrons in a cyclic, delocalized orbital and obeys Hückel's rule for $n = 0$.

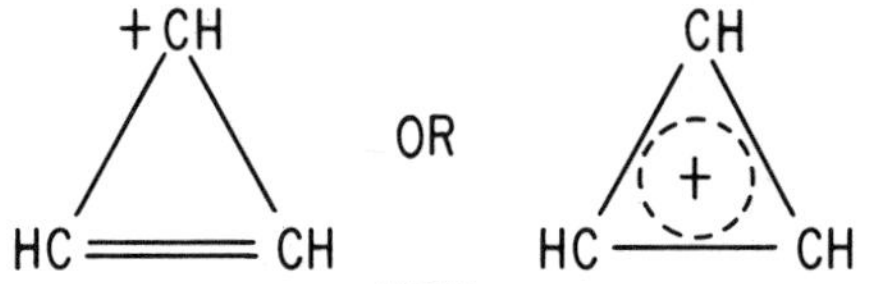

XIII

The delocalization energy of XIII was calculated as the difference in the π-electron energies of XIII and of ethylene and equals 2β. However, the calculations did not take into account other factors, such as ring strain, which are appreciable and which would reduce the stability of the ion. The ring strain of cyclopropene over that of cyclopropane has been given as 27 kcal./mole (*66*). To this ring strain must be added the ring strain of cyclopropane.

In 1957 Breslow reported the first synthesis of the cyclopropenylium system, the tetrafluoroborate salt of the 1,2,3-triphenylcyclopropenyl cation, XV (*7*). Later, other cyclopropenylium compounds were prepared including triaryl-substituted (*10, 44, 48*), diaryl-substituted (*15, 29*), trialkyl- and dialkyl-substituted (*14*), mixed aryl and alkyl-substituted (*14*) and trihalo-substituted compounds (*60*).

Alkyl-substituted cyclopropenylium ions bearing propyl groups are as stable or more stable than phenyl-substituted cyclopropenylium ions.

XIV

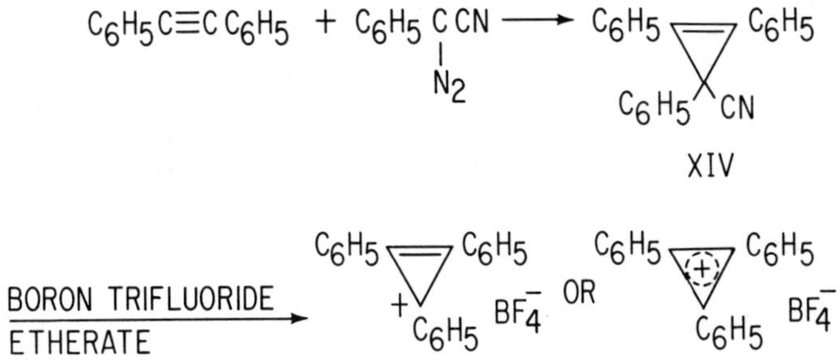

XV

Breslow has suggested that possibly the cyclopropenylium ion is not markedly influenced by resonance effects but is subject to inductive effects (14). Propyl groups would then be better able to disperse the charge than phenyl groups.

Hydride abstraction by the triphenylmethide ion, the method used to prepare tripropylcyclopropenylium ion points to the stability and aromatic character of the cation XVI (14).

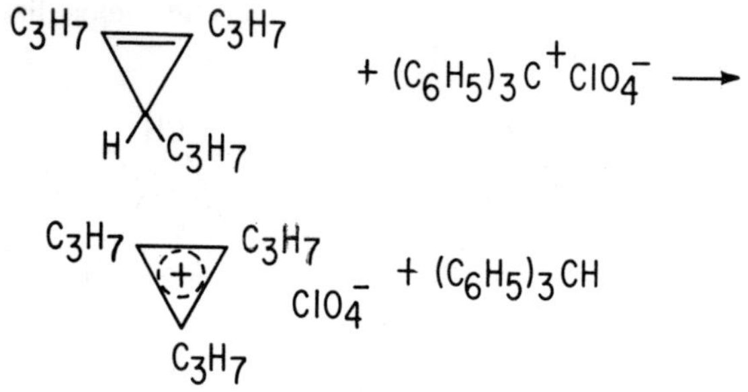

$$C_3H_7 \diagdown \diagup C_3H_7 \quad + (C_6H_5)_3C^+ClO_4^- \longrightarrow$$
$$H \diagdown C_3H_7$$

$$C_3H_7 \diagdown \diagup C_3H_7 \quad + (C_6H_5)_3CH$$
$$\overset{(+)}{\diagdown} \quad ClO_4^-$$
$$C_3H_7$$

XVI

Other evidence for the symmetry and aromatic character of the cyclopropenylium ion was obtained from physical measurements. The cyano compound, XIV, is soluble in nonpolar solvents and gives no precipitate with alcoholic silver nitrate. In contrast, the tetrafluoroborate salt, XV, is a white crystalline solid (m.p. 300°, decomposes) that is insoluble in nonpolar solvents but is soluble in methanol.

An x-ray structural analysis of sym-triphenylcyclopropenylium perchlorate showed that it is made of cyclopropenylium cations and perchlorate anions (57). The carbon-carbon bond distances of the cyclopropenylium ring are 1.376, 1.373, and 1.370 A. Analysis indicated that the three-membered ring and the three phenyl carbon atoms attached to this ring all lie in one plane. The phenyl groups have a propeller-like arrangement about and make angles of 7.6, 12.1, and 21.2° with the central ring.

Spectra studies also indicate an aromatic structure. The absorption peaks of the ultraviolet spectrum of alkyl-substituted cyclopropenylium compounds are below 185 mμ (7). In accordance with predictions of simple Hückel theory, the ultraviolet absorption should occur at relatively short wavelengths since the energy of the π-π^* transition is predicted to be 3β while that of ethylene is 2β. The simplicity of the infrared spectrum of trichlorocyclopropenylium tetrachloroaluminate was taken as evidence of the high symmetry of the trichlorocyclopropenylium ion (60).

The NMR absorption peak for the ring hydrogen of dipropylcyclopropenylium perchlorate is in the region of the spectrum in which aromatic hydrogens absorb rather than in the region in which aliphatic hydrogens absorb. NMR spectroscopy is now the most widely used method of detecting non-benzenoid aromatic systems. The cyclic π-electron system of aromatic molecules can sustain a magnetically induced ring current. Hydrogens on the outside of an aromatic ring are deshielded by the ring current and absorb at lower fields than similar hydrogens not deshielded (38). It is usually assumed that one has an aromatic system if the NMR absorption peak of hydrogens on carbons of the ring system is at a lower field than is expected for the same hydrogens if they were to be considered aliphatic. (The positive charge on the cyclopropenylium ring could also cause a shifting to a lower field.) In addition, the absorption peak for the ring hydrogen of dipropylcyclopropenylium cation has a value in close agreement with the value predicted by the rule of Spiesecke and Schneider (55). This rule states the linear relationship between the chemical shift of aromatic protons and the calculated π-electron density of the aromatic system in question. It has been used frequently as a test for aromatic character.

The sum of the chemical and physical evidence indicates that the cyclopropenylium system does have aromatic properties as predicted by Hückel's calculations.

Hückel molecular orbital calculations for different cyclopropenyl systems indicate that the radical and the anion would have electrons in nonbonding orbitals and that the orbitals are only partially filled (Figure 1). Hückel's rule indicates that neither the cyclopropenyl radical nor

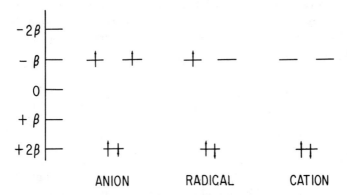

Figure 1.   Hückel molecular orbital energy levels and electron arrangement for different cyclopropenyl systems

the anion should have an appreciable delocalization energy. An ESR study of hexaphenyl-bi-2-cyclopropen-1-yl, XVII, gave no evidence of radicals (13). Thus, there appears to be no appreciable dissociation of XVII into triphenyl-cyclopropenyl radicals.

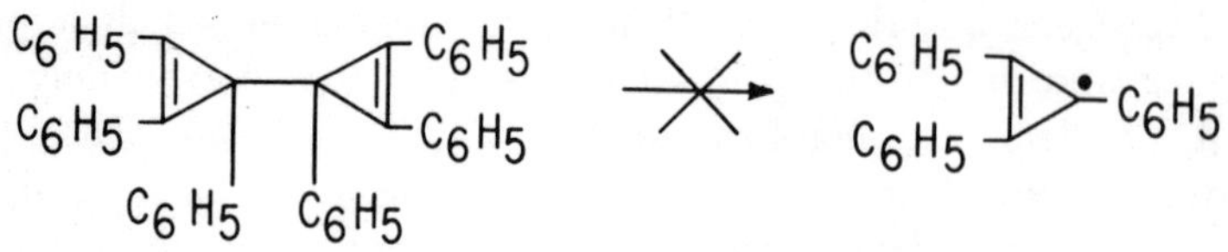

XVII

In other experiments, evidence was obtained that an anion could be formed from XVIII since its reaction with potassium *tert*-butoxide in $(CH_3)_3COD$ led to deuterium exchange for the α-hydrogen (9).

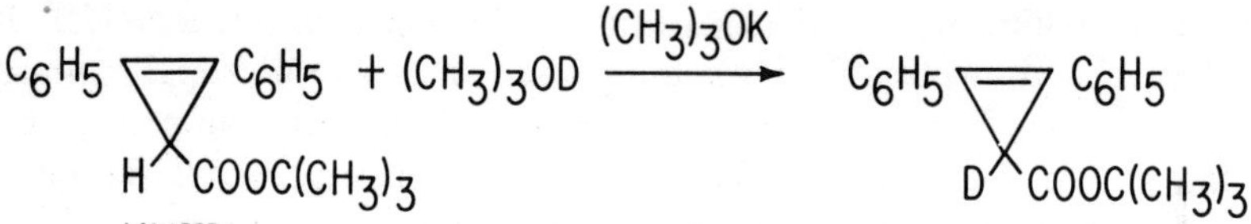

XVIII

However, the exchange with XVIII is much slower than the comparable reaction with a cyclopropane ester and suggests that the double bond has a destabilizing effect on the anion. Anion formation is not a general reaction of cyclopropenes because tritium exchange could not be detected when triphenylcyclopropene was the starting material (11). Thus, neither the anion nor the radical has the relative stability of the cation.

Cyclopropanones have been detected but not isolated. Cyclopropenones, however, have been isolated and are quite stable, from which it can be inferred that the ring of the cyclopropenones has aromatic properties. Resonance theory, as well as molecular orbital calculations, predict aromatic properties for the ring.

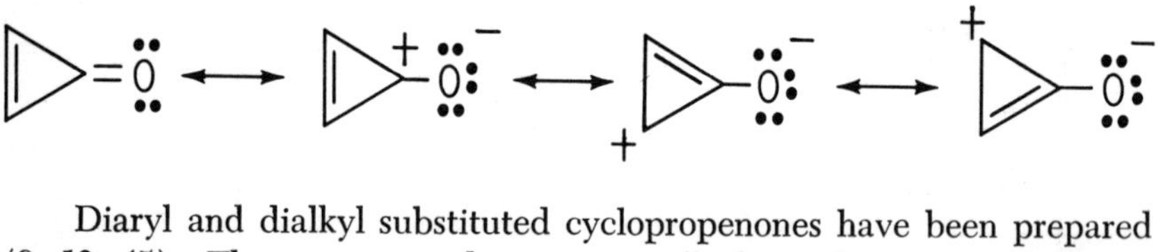

Diaryl and dialkyl substituted cyclopropenones have been prepared (8, 12, 45). These compounds are unusually basic for ketones and can be extracted from carbon tetrachloride by 12N HCl. The basicity of the ketones is related to the presence in the salt of the cyclopropenylium ring and to the absence of the charge separation of the parent ketone.

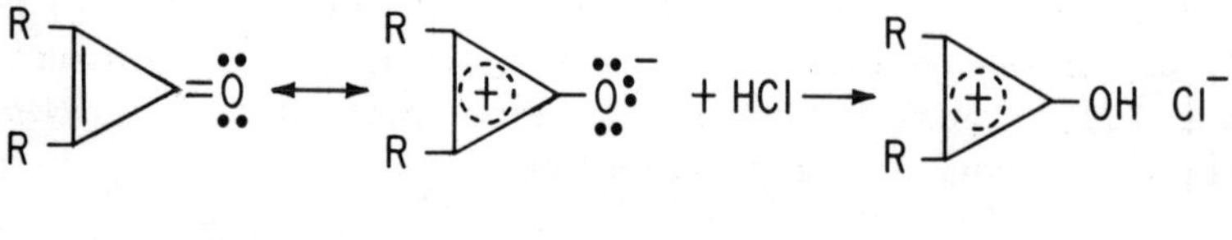

Another compound for which $4n + 2 = 2$ is the dication of cyclo-butene, whose preparation in solution has been reported (*30*). The sub-stituted cyclobutene, XIX, when dissolved in 96% H_2SO_4 forms a deep red solution. The solution possibly yields XX, a dication of cyclobutene.

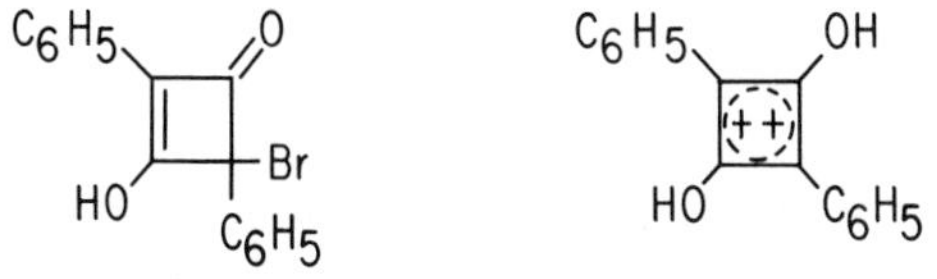

The NMR spectrum of XIX in 96% H_2SO_4 indicates benzylic type carbonium ions are present, which is best explained by the formation of XX.

$4n + 2 = 6$ Systems

Two non-benzenoid aromatic groups of interest which have six π-electrons are the cyclopentadienyl anion and the tropylium ion. For both ions the bonding shells are filled, and Hückel's rule is satisfied. These ions should be relatively stable.

The high acidity, for a hydrocarbon, of cyclopentadiene is attributed to the unusual stability of the cyclopentadienyl anion. [The anion can be formed by the action of potassium in the cold with the parent hydro-carbon (*59*).] Chemists for many years have ascribed aromatic proper-ties to the cation on the basis of this reaction and the fact that the anion has a sextet of π-electrons (*35*). The NMR spectrum of the cation also indicates that the ion is aromatic for the spectrum consists of a single, sharp peak (*31*).

Hückel predicted that the cycloheptatrienylium ion (the tropylium ion), XXI, would be aromatic (*37*).

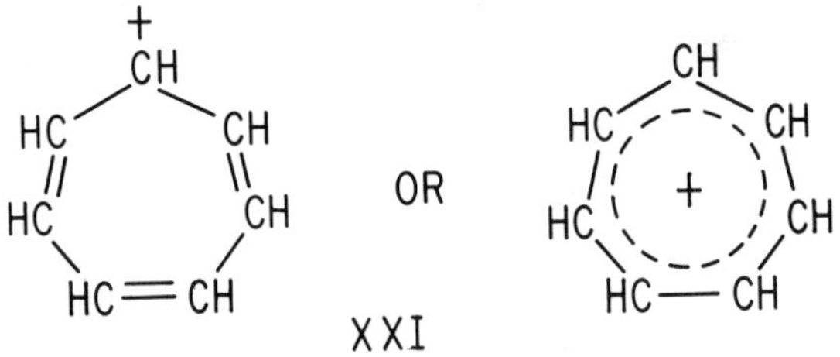

In 1945, Dewar deduced that the seven-membered ring of stipitatic acid could be considered to be a derivative of the tropylium system (*23*). The tropylium ion itself was first prepared in 1954 by Doering and Knox from tropilidene, cycloheptatriene (*27*).

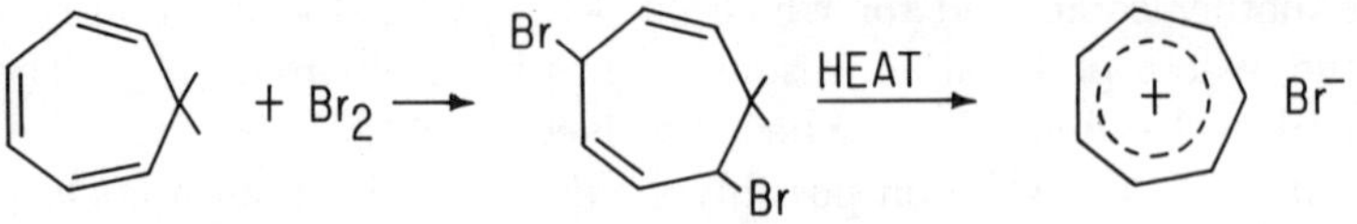

TROPILIDENE

The solubility properties of tropylium bromide (it is insoluble in solvents of low polarity and miscible with water) imply that it is ionic. It has a melting point of 203° and gives an immediate precipitate when silver nitrate is added, indicating that the cation is relatively stable. Its simple infrared spectrum suggests that it is a symmetrical material while its Raman spectrum shows no absorption typical of conjugated, nonaromatic systems. These facts substantiate the conclusion that the tropylium ion is aromatic.

Tropylium bromide is about as acidic as acetic acid when water is the reference base (27).

$$C_7H_7^+ + 2H_2O \rightleftarrows C_7H_7OH + H_3O^+$$

The equilibrium constant for this reaction is approximately 1.8 x 10^{-5}, which implies that the tropylium ion is relatively stable in the presence of the basic water molecules.

Tropylium bromide reacts with many nucleophilic agents to yield products bearing cycloheptatrienyl groups (27), for example:

$$2C_7H_7Br + H_2S \xrightarrow{\text{H}_2\text{O}} (C_7H_7)_2S + 2HBr$$

$$3C_7H_7Br + 4NH_3 \xrightarrow{\text{ethyl ether}} (C_7H_7)_3N + 3NH_4Br$$

These reactions are in contrast to the typical reactions with electrophilic agents of the benzenoid aromatic compounds.

In a reaction analogous to one given by 2,3-diphenylcyclopropenone, tropone (cycloheptatrienone), XXII, forms a salt with hydrogen chloride (22). The salt is stable to such a degree that it can be sublimed. Again,

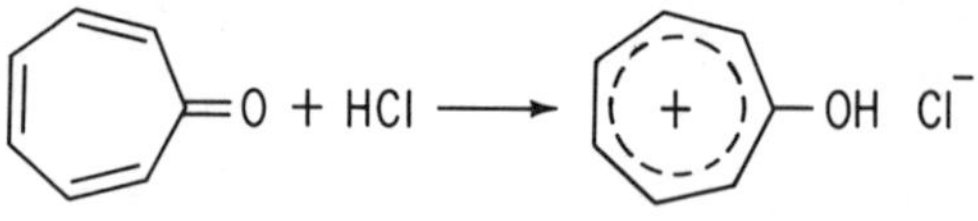

XXII

this unusual stability of the salt of a ketone is probably related to the formation of a non-benzenoid aromatic ring system in the cation (21).

Additional ions which fit Hückel's rule when $n = 1$ are the anion of cyclobutadiene, XXIII, and the dianion of cyclobutene, XXIV. No simple

XXXIII 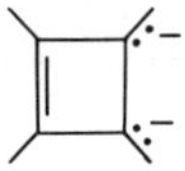   XXXIV

molecules having either of these systems appear to have been prepared to date.

$4n + 2 = 10$ Systems

Recent attempts to prepare the most obvious ring system having 10 π-electrons, cyclodecapentaene, have not been successful. Neither 9,10-dihydronaphthalene nor XXV could be converted to cyclodecapentaene or a derivative of cyclodecapentaene, respectively, under a variety of

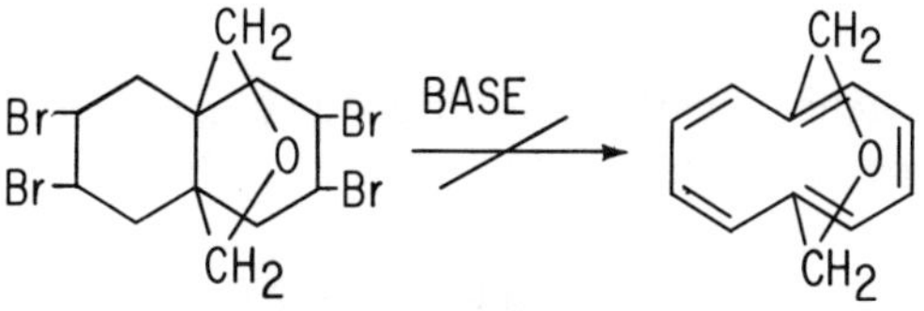

XXV

conditions (6, 61). However, other derivatives of decalin have been converted to compounds which have a cyclodecapentaene ring system (53, 62, 64).

The tetrabromo derivatives XXVI, XXVII, and XXVIII have been converted to bridged undeca-1,3,5,7,9-pentaenes, XXIX, XXX, and XXXI, respectively.

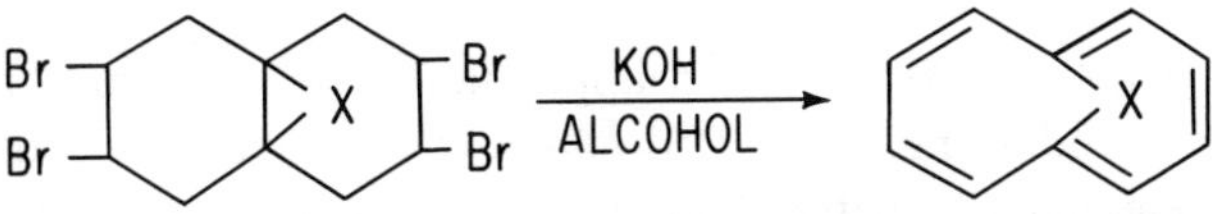

(XXVI) X $=$ CH$_2$ (XXIX) X $=$ CH$_2$
(XXVII) X $=$ O   (XXX) X $=$ O
(XXVIII) X $=$ NHCOCH$_3$ (XXXI) X $=$ NHCOCH$_3$

The products, XXIX, XXX, and XXXI, have 10 π-electrons. NMR and UV spectra of the three compounds point to aromatic structures for these compounds.

Compound XXX undergoes nitration—a typical electrophilic substitution reaction of aromatic compounds—to yield two mononitro compounds

(53). The assignment for the structure of the mononitro product shown was made on the basis of the NMR spectrum. The structure of the

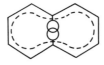

XXX

second nitro compound was not determined. Compound XXIX was also found to undergo nitration, as well as bromination and acetylation, to give substitution products (63).

The fact that cyclodecapentaene itself could not be prepared may be caused by the large nonbonded interactions which would exist be-

tween the hydrogens on carbons 1 and 6. The methylene bridge, oxygen bridge, and nitrogen bridge of XXIX, XXX, and XXXI would remove this interaction.

Ions have been prepared which have 10 π-electrons in a cyclic, delocalized orbital. Reppe and co-workers in 1948 reported that cyclooctatetraene reacts with alkali metals to form dialkali derivatives (49). In 1960 Katz obtained dipotassium cyclooctatetraenide as a solid by causing potassium to react with cyclooctatetraene in tetrahydrofuran (39).

There was no evidence of an appreciable concentration of a radical anion in the product mixture. This fact suggests that the electron affinity of the monoanion is greater than that of the parent hydrocarbon. The NMR spectrum of dipotassium cyclooctatetraenide (and of the dilithium salt) has a single, sharp peak. This observation suggests that the anion is planar and that the resonance energy of the anion must be very large in order to make the cyclooctatetraene ring system planar.

The cyclononatetraenyl anion is a second ion which has a 10-π-electron system. Its preparation was reported simultaneously by two groups of workers (40, 46). Both groups of workers caused a carbene to react with a cyclooctatetraene system to yield a bicyclononatriene.

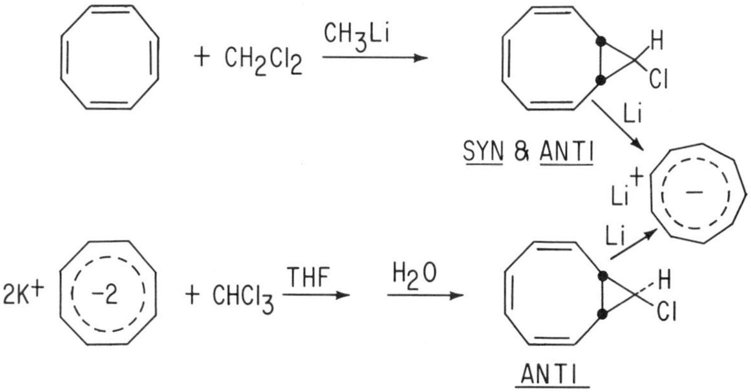

Reaction of lithium with the bicyclononatriene gives the cyclononatetraenyl anion. The lithium salt is sensitive to oxygen and water, but it is stable in an inert atmosphere. The NMR spectrum of the lithium salt has a sharp singlet, whose shape is temperature independent, in the aromatic region. This fact suggests that the anion is planar (47). Cyclononatetraenyllithium can be converted to tetraethylammonium cyclononatetraenide (m.p. 318° decomposes). The infrared and ultraviolet spectra of the tetraethylammonium salt are very simple, as would be expected for a symmetrical group. These experimental facts suggest that the cyclononatetraenyl anion has aromatic properties. It has been calculated that the resonance energy of this anion is 1.5 times that of benzene. (40).

$4n + 2 = > 10$ Systems

Sondheimer and co-workers have prepared a large number of macrocyclic compounds having fully conjugated polyene systems. These compounds are often referred to as annulenes. In the parent compounds, the number of π-electrons equals the number of carbons in the ring. Two of the more intensively studied annulenes are the [14]- and [18]annulenes (macrocyclic polyenes with 14 and 18 carbons). Both of these compounds obey Hückel's rule.

The NMR spectrum at room temperature of the [14]annulene consists of two sharp singlets at τ 4.42 and 3.93 and thus gives no evidence that the compound is aromatic. However, at —60°C., the spectrum consists of peaks at τ 2.4 and 10.0 (33). This observation implies that at low temperature [14]annulene has hydrogens which are highly deshielded and hydrogens which are shielded. The shielding is attributed to the presence of a ring current sustained by an aromatic π-electron system. The four hydrogens which are internal to the ring of [14]annulene, XXXII, are shielded by the ring current and absorb at a high field. The 10 outer hydrogens of [14]annulene absorb at a low field

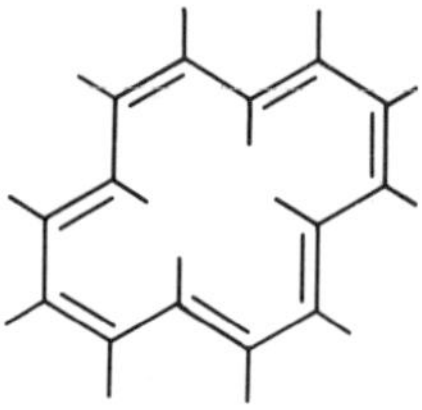

XXXII

in the region characteristic of aromatic hydrogens. At room temperature
there is an equilibrium between different conformations of [14]annulene,
and the NMR absorption peaks coalesce to form a peak at an intermediate
value.

[14]Annulene is not particularly stable (*51*) and gives evidence
that it undergoes addition reactions (*34*).

A monodehydro[14]annulene (XXXIII) and 1,8-bisdehydro[14]-
annulene XXXIV have been prepared (*34*). Both of these compounds
have 14 π-electrons in a cyclic, delocalized orbital (neither the second
pair of π-electrons of the acetylenic linkages nor the center pair of the
cumulene linkage is considered to be in the delocalized orbital). These

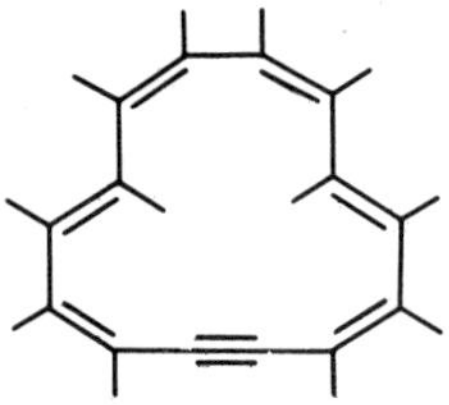

XXXIII 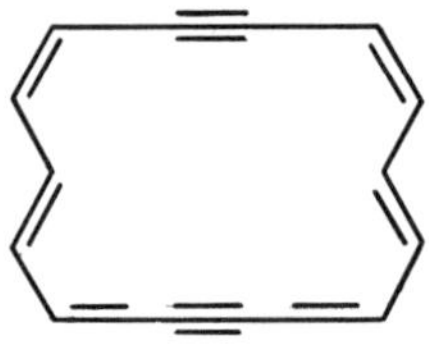   XXXIV

two compounds appear to be aromatic according to their NMR spectra.
Compound XXXIII has a high field band whose areas are in the ratio of
1:5. The high field band is caused by the two internal hydrogens, and
the low field band is caused by the 10 external hydrogens. Both XXXIII
and XXXIV undergo electrophilic substitution reactions (*34*). Treating
XXXIV with cupric nitrate in acetic anhydride at room temperature

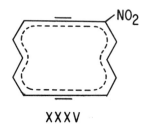

XXXV

yields a mononitro compound. Its NMR spectrum points to XXXV as the structure of the mononitro compound.

Compound XXXIV can also be sulfonated and acetylated to form products whose structures are analogous to XXXV. Compound XXXIII also can be nitrated, sulfonated, and acetylated, but the structures of these products were not completely determined. [14]Annulene, itself, gives no comparable products with these electrophilic agents (*34*).

[16] Annulene has been prepared (*52*). It does not have a number of π-electrons given by Hückel's rule, and thus it would not be expected to be aromatic. It is unstable.

The NMR spectra of [18]annulene, XXXVI, and of tridehydro[18]-annulene, XXXVII, indicate that both have a ring current characteristic of aromatic compounds (*38*).

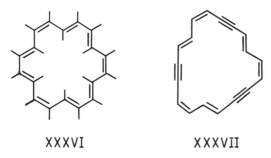

XXXVI   XXXVII

Both XXXVI and XXXVII have 18 π-electrons in a cyclic, delocalized orbital if the second pair of π-electrons of the acetylenic linkages is neglected. The NMR spectrum of XXXVI has a ratio of areas for the aromatic and aliphatic peaks of 2:1, which is that expected for the presence of 10 outer hydrogens and five inner hydrogens. [18]Annulene

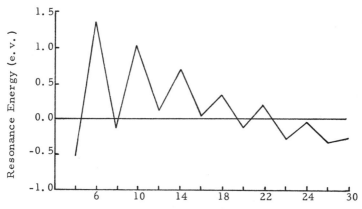

Figure 2. Resonance energies of annulenes as a function of ring size

is reasonably stable; it appears to undergo addition reactions. An x-ray analysis of [18]annulene indicated that all of the carbon-carbon bond lengths are equal and that the molecule is nearly planar (38).

Sondheimer and co-workers also prepared [30]annulene (54). There is no evidence that it is aromatic. It is unstable and undergoes decomposition at room temperature in a benzene solution.

Dewar and Gleicher calculated the resonance energies for the monocyclic, fully conjugated polyenes having 4 to 30 carbons (26). As shown in Figure 2 their calculations indicate that annulenes having more than 22 carbons are not stabilized by resonance. The experimental work of Sondheimer supports this conclusion.

Cyclobutadiene, a compound of interest to many chemists, does not fit Hückel's rule since it has four π-electrons. Many workers attempted to prepare cyclobutadiene but failed. Their failure was caused partly by

$$\underset{\text{Fe(CO)}_3}{\square} \xrightarrow[0°]{Ce^{+4}} [\square] \xrightarrow{HC\equiv CCOOCH_3} \square\text{—COOCH}_3$$

the fact that cyclobutadiene has no delocalization energy (26, 37). Pettit and co-workers recently obtained evidence that cyclobutadiene can be prepared from cyclobutadienyliron tricarbonyl (65).

Cyclobutadiene distills from the reaction flask and is collected at liquid nitrogen temperature. When methyl propiolate is added to the distillate, a Dewar form of methyl benzoate is formed by a Diels-Alder reaction. If no diene is added, the cyclobutadiene dimerizes. This would suggest, in accordance with Hückel's rule, that cyclobutadiene is relatively unstable.

The definition of aromatic character in terms of stability of filled orbitals has, thus, led to the preparation and study of many fascinating, nonclassical aromatic systems.

Literature Cited

(1) Armit, J. W., Robinson, R., *J. Chem. Soc.* **127,** 1604 (1925).
(2) Armstrong, H. E., *J. Chem. Soc.* **51,** 264 (1887).
(3) Baeyer, A., *Ann. Chem.* **245,** 103 (1888).
(4) Bamberger, E., *Ann. Chem.* **257,** 1 (1890); **273,** 373 (1893).
(5) Bamberger, E., *Ber.* **24,** 1758 (1891); **26,** 1946 (1893).
(6) Bloomfield, J. T., Quinlin, W. T., *J. Am. Chem. Soc.* **86,** 2738 (1964).
(7) Breslow, R., *J. Am. Chem. Soc.* **79,** 5318 (1957).
(8) Breslow, R., Altman, L. J., Krebs, A., Mohacsi, E., Murata, I., Peterson, R. A., Posner, J., *J. Am. Chem. Soc.* **87,** 1326 (1965).
(9) Breslow, R., Battiste, M., *Chem. Ind. (London)* 1143 (1958).
(10) Breslow, R., Chang, H. W., *J. Am. Chem. Soc.* **83,** 2367 (1961).

(11) Breslow, R., Dowd, P., *J. Am. Chem. Soc.* **85**, 2729 (1963).
(12) Breslow, R., Eicher, T., Krebs, A., Peterson, R. A., Posner, J., *J. Am. Chem. Soc.* **87**, 1320 (1965).
(13) Breslow, R., Gal, P., *J. Am. Chem. Soc.* **81**, 4747 (1959).
(14) Breslow, R., Höver, H., Chang, H. W., *J. Am. Chem. Soc.* **84**, 3168 (1962).
(15) Breslow, R., Lockhart, J., Chang, H. W., *J. Am. Chem. Soc.* **83**, 2375 (1961).
(16) Breslow, R., Mohacsi, E., *J. Am. Chem. Soc.* **85**, 431 (1963).
(17) Claus, A., *Chem. Ber.* **15**, 1405 (1882).
(18) Claus, A., *J. Prakt. Chem.* **37**, 455 (1888).
(19) Craig, D. P., *J. Chem. Soc.* 3175 (1951).
(20) Craig, D. P., "Theoretical Organic Chemistry," p. 25, Butterworths Scientific Publications, London, 1959.
(21) Culbertson, G., Pettit, R., *J. Am. Chem. Soc.* **85**, 741 (1963).
(22) Dauben, Jr., H. J., Ringold, H. J., *J. Am. Chem. Soc.* **73**, 876 (1951).
(23) Dewar, M. J. S., *Nature* **155**, 50 (1945).
(24) Dewar, M. J. S., *J. Am. Chem. Soc.* **74**, 3345 (1952).
(25) Dewar, M. J. S., Pettit, R., *J. Chem. Soc.* 1617 (1954).
(26) Dewar, M. J. S., Gleicher, G. J., *J. Am. Chem. Soc.* **87**, 685 (1965).
(27) Doering, W. v. E., Knox, L. H., *J. Am. Chem. Soc.* **76**, 3203 (1954); **79**, 352 (1957).
(28) Faraday, M., *Phil. Trans. Roy. Soc., London*, 440 (1825).
(29) Farnum, D. G., Burr, M., *J. Am. Chem. Soc.* **82**, 2651 (1960).
(30) Farnum, D. G., Webster, B., *J. Am. Chem. Soc.* **85**, 3502 (1963).
(31) Fraenkel, G., Carter, R. E., McLachlan, A., Richards, J. H., *J. Am. Chem. Soc.* **82**, 5846 (1960).
(32) Frost, A. A., Musulin, B., *J. Chem. Phys.* **21**, 572 (1953).
(33) Gaoni, Y., Melera, A., Sondheimer, F., Wolovsky, R., *Proc. Chem. Soc.* 397 (1964).
(34) Gaoni, Y., Sondheimer, F., *J. Am. Chem. Soc.* **86**, 521 (1964).
(35) Goss, F. R., Ingold, C. K., *J. Chem. Soc.* 1268 (1928).
(36) Hafner, K., Schneider, J., *Angew. Chem.* **70**, 702 (1958).
(37) Hückel, E., *Z. Physik.* **70**, 204 (1931).
(38) Jackman, L. M., Sondheimer, F., Amiel, Y., Ben-Efraim, D. A., Gaoni, Y., Wolovsky, R., Bothner-By, A. A., *J. Am. Chem. Soc.* **84**, 4307 (1962).
(39) Katz, T. J., *J. Am. Chem. Soc.* **82**, 3784 (1960).
(40) Katz, T. J., Garratt, P. J., *J. Am. Chem. Soc.* **85**, 2852 (1963); **86**, 5194 (1964).
(41) Kekulé, A., *Bull. Soc. Chim. France* [ii], **3**, 98 (1865).
(42) Kekulé, A., *Ann.Chem.* **137**, 129 (1866).
(43) Kekulé, A., *Ann. Chem.* **162**, 77 (1872).
(44) Kende, A. S., *J. Am. Chem. Soc.* **85**, 1882 (1963).
(45) Kursanov, D. N., Volpin, M. E., Koreshkov, Y. D., *Izv. Acad. Nauk SSSR, Otd. Khim. Nauk.* 560 (1959); *Chem. Abstr.* **53**, 21799 (1959).
(46) LaLancette, E. A., Benson, R. E., *J. Am. Chem. Soc.* **85**, 2853 (1963).
(47) LaLancette, E. A., Benson, R. E., *J. Am. Chem. Soc.* **87**, 1941 (1965).
(48) Reid, D. H., Fraser, M., Molloy, B. B., Payne, H. A. S., Sutherland, R. G., *Tetrahedron Letters* 530 (1961).
(49) Reppe, W., Schlichting, O., Klager, K., Toepel, T., *Ann. Chem.* **560**, 1 (1948).
(50) Roberts, J. D., Streitwieser, Jr., A., Regan, C. M., *J. Am. Chem. Soc.* **74**, 4579 (1952).
(51) Sondheimer, F., Gaoni, Y., *J. Am. Chem. Soc.* **82**, 5765 (1960).
(52) Sondheimer, F., Gaoni, Y., *J. Am. Chem. Soc.* **83**, 4863 (1961).
(53) Sondheimer, F., Shani, A., *J. Am. Chem. Soc.* **86**, 3168 (1964).

(54) Sondheimer, F., Wolovsky, R., Amiel, Y., *J. Am. Chem. Soc.* **84**, 274 (1962).
(55) Spiesecke, H., Schneider, W. G., *Tetrahedron Letters* 468 (1961).
(56) Streitwieser, Jr., A., "Molecular Orbital Theory for Organic Chmists," Chap. 2, John Wiley and Sons, New York, 1961.
(57) Sundaralingam, M., Jensen, L. H., *J. Am. Chem. Soc.* **88**, 198 (1966).
(58) Thiele, J., *Ann. Chem.* **306**, 87 (1899).
(59) Thiele, J., *Chem. Ber.* **33**, 666 (1900).
(60) Tobey, S. W., West, R., *J. Am. Chem. Soc.* **86**, 1459 (1964).
(61) Van Tamelen, E. E., Pappas, B., *J. Am. Chem. Soc.* **85**, 3296 (1963).
(62) Vogel, E., Biskup, M., Pretzer, W., Böll, W. A., *Angew, Chem. Intern. Ed. Engl.* **3**, 642 (1964).
(63) Vogel, E., Böll, W. A., *Angew. Chem. Intern. Ed. Engl.* **3**, 642 (1964).
(64) Vogel, E., Roth, H. D., *Angew. Chem. Intern. Ed. Engl.* **3**, 228 (1964).
(65) Watts, L., Fitzpatrick, J. D., Pettit, R., *J. Am. Chem. Soc.* **87**, 3253 (1965).
(66) Wiberg, K. B., Bartley, W. J., Lossing, F. P., *J. Am. Chem. Soc.* **84**, 3980 (1962).

RECEIVED December 14, 1965.

The Chemical Prehistory of the Tetrahedron, Octahedron, Icosahedron, and Hexagon

O. T. BENFEY and LEWIS FIKES[1]

Earlham College, Richmond, Ind.

Geometry has played a variety of roles in explaining natural phenomena at different periods of history. The discovery of the limited number of regular solids by the Pythagoreans was followed soon by the linking of the five solids to Empedocles' four elements plus the fifth essence associated with the heavens. Plato's "Timaeus" develops a quantitative atomism based on these figures and their constituent triangles—an atomism in many respects closer to modern thought than that of Democritus. The regular solids reappear in Renaissance discussions of natural phenomena. With Dalton's atomism, spatial arrangements of the atoms were considered early, and eventually the regular solids were incorporated into modern structural theory.

Kekulé's choice of the hexagon as the geometric pattern for benzene was not as novel as it first seemed. Laurent had introduced the hexagon into discussions of organic chemistry, though in a quite different context. At the time Kekulé proposed it, he also proposed an alternate—a triangular structure for benzene, with three carbons at apexes and three at the midpoints of sides (20, 21). Kekulé's use of regular geometric figures was a considerable deviation from his earlier "sausage formulas," which had carefully avoided all geometric or spatial implications.

This paper traces the role of certain regular geometric figures, which were used to explain the structures and properties of matter, in the hope of pointing out some of the transformations that the relationship between geometry and chemistry has undergone from classic to modern times.

[1] Present address: Department of Chemistry, Ohio State University, Columbus, Ohio.

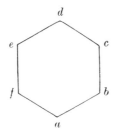
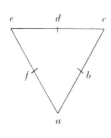

The fact that only few regular geometric solids can be conceived must have profoundly affected the ancient thinkers, for soon after this discovery, the solids were related to the structure of matter and the universe. Initially only four regular polyhedra were known, three of which were constructed of equilateral triangles: the tetrahedron, octahedron, and icosahedron, containing four, eight, and twenty sides, respectively. The fourth polyhedron was the cube—constructed of squares. By regular polyhedron is meant a solid, all of whose angles, edges, and faces are identical. It is easy to demonstrate that not more than three finite regular solids composed of equilateral triangles can exist.

To have a solid angle, at least three edges must meet at a point. If three equal-length edges do meet at every apex and if they join each other in triangular faces, the solid is a tetrahedron. If, instead, four

Tetrahedron

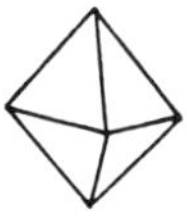

Octahedron

equilateral triangles meet at each apex, we obtain an octahedron while with five, an icosahedron results. If six equilateral triangles meet at a point, we obtain the endless two-dimensional lattice, containing the familiar hexagon structure. Seven equilateral triangles meet at a point only with puckering and hence cannot give a regular solid. The plane hexagonal pattern can be thought of as a closed surface with an infinite

Icosahedron

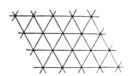

Hexagonal Plane
Triangular Tesselation

number of sides and hence can be included among the regular solids. This was first recognized by Kepler, and the infinite pattern is known among mathematicians as a tesselation.

To proceed with the roster of regular solids, square instead of triangular faces may be used. If three edges meet, the cube results; with four, the checkerboard plane.

Cube   Square Tesselation

Regular pentagons which meet three to a point yield the dodecahedron; they cannot be arranged to form a tesselation. Hexagons, with 120° angles, can only form a tesselation, and any regular polygon with more than six sides has angles larger than 120° and hence cannot meet with two others to form a solid angle at all.

Dodecahedron   Hexagonal Tesselation

The regular geometric shapes, then, include only five finite polyhedra, whose geometric characteristics are given in Table I, and three plane tesselations, based on the triangle, square, and hexagon respectively.

Table I. The Regular Solids

	Face	Number of edges meeting at each vertex	Number of edges	Number of faces	Number of vertices
Tetrahedron	triangle	3	6	4	4
Octahedron	triangle	4	12	8	6
Cube	square	3	12	6	8
Dodecahedron	pentagon	3	30	12	20
Icosahedron	triangle	5	30	20	12

One of the remarkable properties of these solids is summarized by the Descartes-Euler formula:

Number of vertices + Number of faces = Number of edges + 2

Coxeter (5) has remarked that to ask who first constructed the five regular solids is probably as futile as to ask who first used fire. Their early history is lost in antiquity. Because Plato (—427 to —347) discusses them at length in the "Timaeus," they are often known as the Platonic solids. The Pythagoreans knew them and could construct them by putting together triangles, squares, and pentagons made of leather, cloth, or parchment. (The notation —427 to —347 is used instead of 427–347 B.C., following the suggestion of J. Needham (32) as being less cumbersome and more suitable to a world intellectual community.)

At this period the concept of matter was the four-element theory of Empedocles—all matter being described in terms of the different contributions of earth, air, fire, and water. Philolaos of Tarentum (ca.—430) felt the need of a fifth element because he was convinced that there must be a connection between the elements and the regular solids. He is believed to have conceived of it as the unobservable substance, the "apeiron" of Anaximander, of which all visible substances are made. Plato and Aristotle identified the fifth solid with the sphere of the stars or the substance of the heavenly bodies, thus banishing it from the earth and the realm below the moon (12, 31).

Plato's "Timaeus" contains a detailed account for constructing four of the regular solids from two types of triangles, identifying the four solids with the four elements, and describing the materials of nature— metals, ice, water, steam, oils, juices, rocks, etc.—in terms of the elements and their geometric constituents.

The Pythagoreans and Plato

Why was there such an interest in explaining nature in terms of geometry? The answer may well be that it was seen as a way out of the Pythagorean crisis. The Pythagoreans, amazed by the properties of integral numbers and their relation to physical phenomena, such as the pitch of strings of different length and hammers of different weight, had resolved to construct the universe on the basis of numbers alone. However, the length of the diagonal of a square of unit length could not be expressed as the ratio of integers. Whatever ratio was chosen, did not, when squared, give the number 2 exactly.

The religious mysticism based on numbers collapsed. It is said that the Pythagoreans tried to keep the scandal of irrationality secret and even banished or killed one member who had talked too much.

Plato constructed his universe on triangles rather than numbers and chose two triangles that involved the square roots of 2 and 3, respectively —the half-square with sides $1:1:\sqrt{2}$ and the half-equilateral triangle with sides $1:2:\sqrt{3}$. Thereby, the scandal was quieted and irrational numbers were incorporated into the essence of nature. Karl Popper has gone even further and suggested that Plato hoped all irrational numbers were perhaps sums or multiples of $\sqrt{2}$ and $\sqrt{3}$. Their sum, for instance, comes to 3.146 . . . a figure as close as the Greeks' best estimate for that other irrational number, π (43).

Popper suggested that the $1:1:\sqrt{2}$ and $1:2:\sqrt{3}$ triangles were chosen from all the possible ones in order to incorporate the two square roots, but there is a simpler explanation more directly in line with the text of the "Timaeus." Plato's question was the following (37):

What are the most perfect bodies that can be constructed, four in number, unlike one another, but such that some can be generated out of one another by resolution? If we can hit upon the answer to this, we have the truth concerning the generation of earth and fire and of the bodies which stand as proportionals between them.

The bodies that stand as "proportionals" between earth and fire were identified earlier by Plato as water and air.

Given the need for only four bodies and for the interconvertibility of some of them, the four Platonic solids made of triangles and squares were obvious candidates. Three of them—the tetrahedron, octahedron, and icosahedron—could be conceived of as interconvertible since they were all constructed of the same type of triangle. Transmutation was therefore explainable.

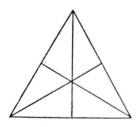
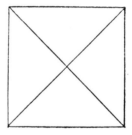

Plato's Triangle   Plato's Square

For reasons not apparent, Plato divided the equilateral triangle into six parts to obtain his unit building block, the $1:2:\sqrt{3}$ triangle. Division into two would have been sufficient but perhaps was not considered as symmetrical. Similarly, Plato used the quarter-square rather than the half-square for his $1:1:\sqrt{2}$ triangle.

From the first type of triangle he proceeded to construct the tetrahedron, octahedron, and icosahedron, and from the second the cube. The crucial step then was to assign the correct elements to the four figures. With excellent premonition the tetrahedron was taken as belonging to fire because of its smallness (smallest number of sides), its sharpest corners, and its presumed high mobility. The cube was assigned to earth as being suitable for a stable substance. Air, being more mobile and lighter than water, was assigned the octahedron while water was left with the icosahedron.

Plato believed that the constituent triangles and the single geometric figures were too small to be seen; only aggregates were visible. For Plato, already, properties on the macroscopic level paralleled those of the "atomic units" and could be used to deduce details about those units. A parallel may be drawn with the period around 1880 when the question arose whether the tetrahedral bonding of carbon was caused by a tetrahedral shape of its atom or by repulsive forces operating around a spherical atom. The latter view was championed by Alfred Werner (13).

Using geometric figures Plato could develop not only a qualitative but a quantitative picture of transmutation. Since the tetrahedron has four, the octahedron eight, and the icosahedron 20 sides, the following transformations were envisaged:

$$
\begin{array}{llll}
\text{1 Water} & \rightarrow & \text{1 Fire} & + & \text{2 Air} & \quad (1) \\
\text{icosahedron (20)} & & \text{tetrahedron (4)} & & \text{octahedra (2 x 8)} \\
\text{1 Air} & \rightarrow & \text{2 Fire} & & & \quad (2) \\
\text{octahedron (8)} & & \text{tetrahedra (2 x 4)} \\
\text{5 Air} & \rightarrow & \text{2 Water} & & & \quad (3) \\
\text{octahedra (5 x 8)} & & \text{icosahedra (2 x 20)}
\end{array}
$$

Plato even talks about reversibility. He suggests that under certain conditions Equation 2 can be reversed, two fire particles reconstituting one air particle.

The fourth element, earth, could not be involved in transmutation; it could only be decomposed into its triangles and reconstituted. Finally, the primary triangles existed in different sizes, thus accounting reasonably well for the filling of all space and making it easier to account for the diversity of material substances.

Scientists today usually idealize Democritus and Leucippus as the forerunners of modern atomism. However, in some respects the concepts of Plato are closer to our views. Democritus conceived of an unlimited number of shapes, Plato of a severely limited number. Democritus denied transmutation; Plato permitted it of his geometric bodies by allowing them to disintegrate into simpler parts which were not transmutable. Admittedly, Plato did not permit a vacuum while Democritus did. Quantum theory, however, can accommodate either view regarding vacua. Finally, mathematical theorems controlled the behavior of Plato's building blocks while Democritus' atoms were only qualitatively described (44).

According to Ihde (15), the "Timaeus" was the only dialogue of Plato to be translated into Latin before the Middle Ages and considerably influenced European ideas of nature. However, the crucial passages concerning the regular solids were not translated at that time (39).

Aristotelian works were largely discovered through Spain—via the Arabic transmission of Greek knowledge—and soon dominated western philosophical thought in the Middle Ages. However, the Arabs had also maintained the Platonic-Pythagorean interest in mathematics and had fused this attitude with Aristotelian ideas. Much of the "modern" concept of nature as something to be understood mathematically and bent to human ends stems from Arabic practice. The Arabs carried further one aspect of the mathematics of regular polyhedra—the study of tesselations particularly on a spherical surface. These were described by Abû'l Wafâ (940–998) (6).

The Middle Ages

There is little evidence that the Platonic solids played a significant role in natural speculation during the Milddle Ages. In natural philosophy we meet them again in the works of Offusius, Davidson, and Kepler.

Plato's theory seems to have influenced Offusius, an astronomer and astrologer active in the 16th century. Tycho Brahe wrote approvingly of his work. Like Plato, Offusius believed that nature followed a certain numerical and geometrical order, and accordingly he used the five regular solids to try to "explain" this order. However, he did not identify the four elements with four of the five regular solids as Plato had done. Instead, four of the solids were identified with the four qualities—hot, cold, dry, wet, which Aristotle had considered more fundamental than the elements—while the other regular solid (the dodecahedron) was identified with the sphere of the fixed stars. Offusius' scheme is as follows:

hot	:	dry	::	pyramid	:	cube
hot	:	cold	::	pyramid	:	octahedron
cold	:	wet	::	octahedron	:	icosahedron

In trying to relate these same four regular solids to the order that is observed in the heavens, Offusius engaged in mystic calculations of the type later indulged in by Kepler. Offusius proposed that:

pyramid	:	square	::	28 1/4	:	84 3/4
octahedron	:	icosahedron	::	113 391/512	:	133 121/512

These numbers total 360, a divine proportion according to Offusius and undiscovered by anyone before him as far as he knew. He omitted the dodecahedron in these calculations because he thought it corresponded excellently to the sphere of the fixed stars—the dodecahedron being almost spherical (42).

Plato's ideas about the relationship between the regular solids and the elements were adopted almost entirely by the iatrochemist, William

Davidson [also spelled Davison, Davissone, 1593–1669]. Davidson said that the Platonic doctrine was sufficient to explain the true cause of the different forms, shapes, and proportions of the bodies found in nature. He also tried to combine geometric ideas associated with the Platonic bodies with the three principles of alchemy, saying: "as a solid angle cannot be made without three planes, so a natural body cannot be made without salt, sulfur, and mercury" (38). Born in Aberdeen, Scotland, Davidson held the first chair of chemistry to be founded in Paris. He was also physician to Louis XIII of France (10).

One of two engraved plates in Davidson's "Les Elemens de la Philosophie de l'Art du Feu ou Chemie" (9) shows 20 geometric solid forms, headed by the Platonic polyhedra. The second plate (Figure 1) was

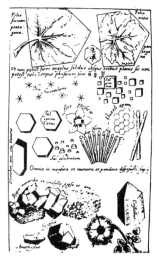

Figure 1. Plate from "Les Elemens de la Philosophie de l'Art du Feu ou Chemie" (9), showing natural forms with shapes similar to those of regular solids

intended to illustrate his claim that these forms could explain all natural forms found in nature, not only of crystals but also of materials of the plant and animal realm. One figure of a section of a honeycomb looks much like the formula of a condensed aromatic hydrocarbon. The hexagon also appears enclosing a bee, as the form of "hexagonal snow" (*nix sexangularis*), and of the "so-called carbonate of ammonia" (*sal cornu Cerui*) (38). In the middle of the plate appears the statement "Thou hast ordered all things in measure, number, and weight."

Johannes Kepler's use of the Platonic bodies in constructing his model of the universe is well known, and it appears that he was original in using these solids to explain astronomical phenomena. His discovery occurred on July 9, 1595. While lecturing, he found it necessary to draw a figure in which an outer circle was circumscribed about a triangle in which an inner circle was inscribed (24). As he looked at the two circles, it suddenly struck him that their ratios were the same as those of the orbits of Saturn and Jupiter. He then reasoned that the universe is built around certain symmetrical figures—triangle, square, pentagon, etc.—which form its invisible skeleton. However, when he tried to inscribe a square between Jupiter and Mars, a pentagon between Mars and Earth, and a hexagon between Earth and Venus, he found that the scheme did not work. Nevertheless, feeling that he was close to the secret, he decided to try three-dimensional forms instead of two-dimensional ones, and it worked, or at least so he thought. Kepler's "Harmony of the Spheres" is mainly a discussion of the Platonic solids, their relation to the four elements and to the planets.

In dealing with the problem of the relative distances of the planets, Kepler argued that the five intervals between the six planetary spheres

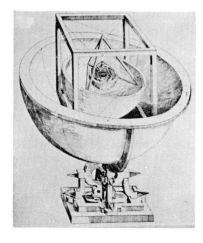

Figure 2. Planetary orbits embedded in the regular solids, from Kepler's "Mysterium Cosmographicum" (23)

of Copernicus could be filled with the five regular solids, so that the sphere of Saturn circumscribes a cube in which the sphere of Jupiter is inscribed; the latter circumscribes a tetrahedron in which the sphere of Mars is inscribed. Next is the dodecahedron, the sphere of Earth, the icosahedron, the sphere of Venus, the octahedron, and the sphere of Mercury (Figure 2) (23). Kepler believed he had thus penetrated the secrets of the Creator.

The regular figures of geometry begin to appear again almost as soon as chemistry was recast in its modern conceptual form by Lavoisier and Dalton. While parts of Dalton's "New System of Chemical Philosophy" were still being published, Wollaston was pointing out that numerical relationships were not enough—geometric relationships were required to explain chemical behavior (46):

. . . when our views are sufficiently extended to enable us to reason with precision concerning the proportions of elementary atoms, we shall find the arithmetical relation alone will not be sufficient to explain their mutual action and that we shall be obliged to acquire a geometrical conception of their relative arrangement in all the three dimensions of solid extension.

For instance, if we suppose the limit to the approach of particles to be the same in all directions, and hence their virtual extent to be spherical (which is the most simple hypothesis), in this case, when different sorts combine singly there is but one mode of union. If they unite in the proportion of two to one, the two particles will naturally arrange themselves at opposite poles of that to which they unite. If there be three, they might be arranged with regularity, at the angles of an equilateral triangle in a great circle surrounding the single spherule; but in this arrangement, for want of similar matter at the poles of this circle, the equilibrium would be unstable, and would be liable to be deranged by the slightest force of adjacent combinations; but when the number of one set of particles exceeds in the proportion of four to one, then, on the contrary, a stable equilibrium may again take place, if the four particles are situated at the angles of the four equilateral triangles composing a regular tetrahedron.

In 1811, in answer to criticisms by Bostock (1), Dalton himself pursued Wollaston's line of thought but curiously confined himself to two dimensions. For him AB_4 would be a square-planar, not a tetrahedral structure (7):

When an element A has an affinity for another B, I see no mechanical reason why it should not take as many atoms of B as are presented to it and can possibly come into contact with it (which may probably be 12 in general), except so far as the repulsion of the atoms of B among themselves are more than a match for the attraction of an atom of A. Now this repulsion begins with 2 atoms of B to one of A, in which case the 2 atoms of B are diametrically opposed; it increases with 3 atoms of B to one of A, in which case the atoms of B are only 120° asunder; with 4 atoms of B it is still greater as the distance is then only 90°; and so on in proportion to the number of atoms.

Soon he must have extended his conceptions to three dimension, for in 1842 in a paper on acids, bases, and water which he published privately, he describes the models his friend Ewart constructed for him 30 years earlier from spheres and pins. The AB_4 model is still square planar, but AB_6 is octahedral, and AB_5 a triangular bipyramid (8).

Wollaston's and Dalton's ideas had little effect. Chemists were skeptical about the possibility of determining the relative positions of atoms in space, a skepticism which reached its peak in Kolbe's attack on van't Hoff (25).

The Tetrahedron Concept

Pasteur was the first to apply the tetrahedron concept to organic chemistry when he summarized his studies on optical rotation in solution (33):

We know, in fact, on the one hand, that the molecular arrangements of the tartaric acids are dissymmetric, and on the other that they are rigorously the same, with the sole difference of presenting dissymmetries in opposite directions. Are the atoms of the right [clockwise rotating] acid grouped according to the spire of a dextrorse [right-handed] helix, or placed at the summits of an irregular tetrahedron, or disposed according to such or such determined dissymmetric assemblage? We are unable to reply to these questions. But what cannot be doubted is, that there is a grouping of atoms according to an order dissymmetric to a non-superposable image. What is not less certain is, that the atoms of the left acid precisely realize the inverse dissymmetric grouping of this one.

The tetrahedron concept was first applied to carbon by Butlerov in 1862 (2). In an attempt to explain the assumed isomerism of ethyl hydride, $C_2H_5 \cdot H$, and dimethyl, $CH_3 \cdot CH_3$, he suggested a model of a tetrahedral carbon atom, with each face capable of attaching a univalent atom or group. He then calculated the number of isomers to be expected in the case of methane and some of its derivatives if two, three, or four of the valences of carbon (even when all were bonded to hydrogen) were different in character. By assuming differences in carbon affinities, he was able to explain the "isomerism" between the two hydrocarbons.

Five years later Kekulé described a tetrahedral carbon model (22) useful for visualizing the links in acetylene, $H—C≡C—H$ and hydrogen cyanide, $H—C≡N$. Recently, Gillis discovered a copy of Butlerov's earlier article with annotations by Kekulé (11). It seems almost certain, therefore, that Butlerov's paper played a role in the development of Kekulé's models. van't Hoff worked with Kekulé in 1872 and published his own stereochemical views two years later. Both van't Hoff and le Bel worked in Wurtz's laboratory in 1873, but they do not seem to have discussed the question. van't Hoff was influenced by the statement by Johannes Adolf Wislicenus (1835–1902) (45) that the explanation for optical activity in the lactic acids must be sought in the spatial relationships of the constituent atoms while le Bel tended to follow Pasteur's more abstract line of reasoning regarding symmetry properties.

With regard to the occurrence of the hexagon in structural discussions, it appeared among Dalton's molecule diagrams with no spatial significance and again as a true geometric hexagon figure in a dicussion by Laurent of organic substitution (26):

To render the reciprocal replacement of the two residues intelligible, I will suppose, that in ammonia and chloride of benzoyl the atoms are arranged so as to form hexagonal figures.

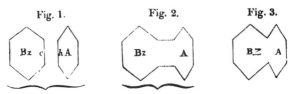

Bz and A, Fig. 1, represent chloride of benzoyl and ammonia at the moment when they react upon one another, the face c being opposed to the face h, which is to be set free. Bz and A, Fig. 2, represent the two residues during the reaction, and Bz and A, Fig. 3, the two residues after the reaction having reciprocally filled up the two voids formed in A and Bz by the removal of the chlorine and hydrogen "faces."

Kekulé reproduced this set of hexagonal figures together with Laurent's explanation in his famous paper on the tetravalence and chain-forming capacity of carbon atoms (19). We are certain, therefore, that Kekulé had seen a hexagonal figure in chemical discussions.

Laurent invariably thought geometrically, presumably owing to his early crystallographic training. In his doctoral dissertation on organic substitution theory he explained his ideas by a diagram of a rectangular prism (an elongated cube), whose corners were occupied by carbon atoms, with hydrogen atoms at the midpoints of the edges (16). Most likely Pasteur's interest in crystallography, which finally led to his work on molecular asymmetry, was also kindled by Laurent, whom he assisted for a while in crystallographic experiments.

The resurgence of geometric ideas then came rapidly. After the geometric solution of the benzene problem (1865) and the tetrahedral carbon atom (1874), Alfred Werner directed analogous lines of thought to the realm of inorganic complex compounds, and made the octahedron (1891) the key to much of that field. The octahedron had been suggested some years earlier as a possible structure for benzene (40), and has reappeared recently in discussions of certain boron hydrides (see below).

The Icosahedron

The icosahedron followed much later and did not find a significant niche in chemistry until the understanding of boron and the boron hydrides was organized around the icosahedral concept.

That story begins in 1941 with a paper by two Russians, Zhdanov and Sevast'yanov (48) on the x-ray diffraction analysis of crystalline boron carbide B_4C (now recognized as $B_{12}C_3$). They suggested that the structure can be formally considered as similar to an NaCl lattice, with a compact group of twelve boron atoms at the Na^+ sites and a linear group of three carbon atoms at the Cl^- sites. The authors, in addition, gave detailed locations for each atom.

This paper was not abstracted in *Chemical Abstracts* until April 1943, whereupon it came to the attention of Clark and Hoard (4) of Cornell University, who were also engaged in an x-ray study of B_4C. E. J. Crane of *Chemical Abstracts* sent the Cornell chemists a copy of the original Russian article, in which they found that their own partial results agreed completely with those published. Nine of 15 atoms in the unit of structure had been definitely placed when the abstract appeared. The remaining positions were subsequently confirmed by the Cornell group. A model was then constructed. In the words of the authors (4):

Closer inspection, entailing the construction of a rough model [Figure 3] reveals remarkably enough, that the twelve boron atoms of the group are arranged at the vertices of a nearly regular icosahedron. The

Journal of the American Chemical Society

Figure 3.   Rough model of the boron carbide structure of Clark and Hoard (4), showing an icosahedral cluster of 12 boron atoms

regular icosahedron is one of the five possible regular solids. It has twenty equilateral triangular faces, twelve vertices, and thirty edges. There are six five-fold axes, ten three-fold axes, fifteen two-fold axes, and fifteen planes of symmetry. Each boron atom has six-fold coordination, being bonded to five others in the icosahedral group and to one other atom outside so as to lie approximately at the center of a pentagonal pyramid.

The photograph of the boron carbide model included in the article almost certainly represents the first appearance of an icosahedron in chemical discussions since alchemical times.

The next appearance of the icosahedron occurs in the search for a satisfactory structure of decaborane, $B_{10}H_{14}$. Silbiger and Bauer (41) attempted to fit their new electron diffraction data to numerous structures without success. In a footnote to their paper added in proof, they mention new findings by Kasper, Lucht, and Harker based on x-ray diffraction studies and comment: "Due to its unexpected form, their configuration had not been considered in the above electron diffraction study." Silbiger and Bauer then gave a complicated verbal description of the newly proposed structure.

Kasper, Lucht, and Harker reported their findings later the same year (18), including a diagram of the $B_{10}H_{14}$ structure clearly showing the boron atoms to be located at 10 of the vertices of an icosahedron. However, the icosahedron was not mentioned until the detailed crystallographic data were published in 1950 (17) and further commented on in 1951 (29).

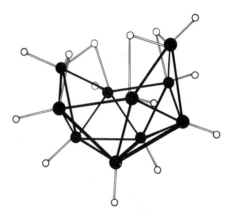

Journal of the American Chemical Society

Figure 4. The decaborane structure after Kasper, Lucht, and Harker (18)

The demonstration of the icosahedral arrangement of the atoms of decaborane was a major breakthrough in boron hydride chemistry because it soon became apparent that most boron hydrides were hydrogenated fragments of boron icosahedra. This was proposed by Lipscomb (27), who also suggested fragmented octahedral structures for a few boron hydrides.

A remarkably stable icosahedral ion, $B_{12}H_{12}^{-2}$, was predicted by Longuet-Higgins and Roberts in 1955 (28) and prepared by Pitochelli

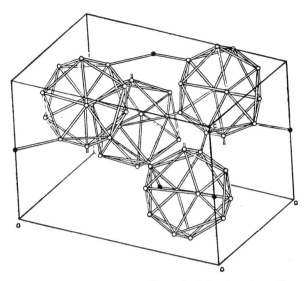

Journal of the American Chemical Society

*Figure 5. Diagram of structure of
tetragonal boron (14)*

and Hawthorne in 1960 (*36*). Its icosahedral structure was determined
the same year (*47*).

The icosahedral structure of the solid element, boron, was recognized
in 1958. First an unusual crystalline modification of boron was shown to
consist of linked icosahedra with a structure "essentially the same as the

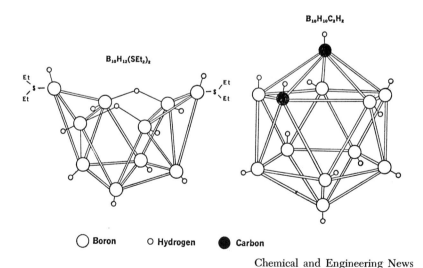

Chemical and Engineering News

Figure 6. Structure of carborane, $B_{10}C_2H_{12}$ (3)

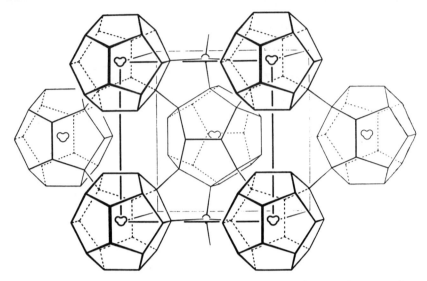

Courtesy Pergamon Press

Figure 7.   Structure of water proposed by Pauling (35)

boron carbide" (*30*). Later that year, other forms of crystalline boron were shown to consist of various arrangements of boron icosahedra (Figure 5) (*14*).

The icosahedral arrangement is so stable that the incomplete icosahedral structure of decaborane will react (under certain conditions) with an acetylene derivative, producing a completed icosahedron, two of whose vertices are carbon atoms. The resulting carborane, $B_{10}C_2H_{12}$, has a hydrogen atom bonded to each vertex atom (Figure 6). Thus, the carbon atoms show the unusual coordination number of 6, dictated by the stability of the icosahedron.

Finally, the dodecahedron was hauled down from the celestial spheres to serve as the structure of certain hydrate clathrates and possibly of certain regions of water (Figure 7) (*35*).

Having learned to use the Platonic solids as explanatory devices, we are again discovering the remarkably small number of regular solids. The first book to organize structural chemistry according to the five regular solids has recently appeared (*34*).

Literature Cited

(1) Bostock, J., *Nicholson's J.* **28**, 280 (1811).
(2) Butlerov, A. M., *Z. Chem. Pharm.* **5**, 297 (1862).
(3) *Chem. Eng. News* **41**, 62 (Dec. 9, 1963).
(4) Clark, H. K., Hoard, J. L., *J. Am. Chem. Soc.* **65**, 2115 (1943).
(5) Coxeter, H. S. M., "Regular Polytopes," p. 13, Pitman Publishing Corp., New York, 1948.

(6) *Ibid.*, p. 73.
(7) Dalton, J., *Nicholson's J.* **29**, 147 (1811). Quoted in Palmer, W. G., "A History of the Concept of Valence to 1930," p. 8, Cambridge University Press, Cambridge, 1965.
(8) Dalton, J., "On Phosphates and Arsenates; Acids, Bases, and Water; A New Method of Analyzing Sugar," Harrison, Manchester, 1842. Quoted in Palmer, W. G., *op. cit.*, p. 12.
(9) Davidson, W., "Les Elements de la Philosophie de l'Art du Feu ou Chemie," F. Piot, Paris, 1651, translation of a Latin version of 1633-35.
(10) DeMilt, C. M., *J. Chem. Ed.* **18**, 503 (1941).
(11) Gillis, J., *Mededel. Koninkl. Vlaam. Acad. Wetenschapp.* **20**, 3 (1958).
(12) Heath, T., "A History of Greek Mathematics," Vol. I, pp. 158-162, Oxford University Press, Oxford, 1921.
(13) Henrich, F., "Theories of Organic Chemistry," p. 75, Wiley and Sons, New York, 1922.
(14) Hoard, J. L., Hughes, R. E., Sands, D. E., *J. Am. Chem. Soc.* **80**, 4507 (1958).
(15) Ihde, A. J., "The Development of Modern Chemistry," p. 8, Harper and Row, New York, 1964.
(16) *Ibid.*, p. 195.
(17) Kasper, J. S., Lucht, C. M., Harker, D., *Acta Cryst.* **3**, 436 (1950).
(18) Kasper, J. S., Lucht, C. M., Harker, D., *J. Am. Chem. Soc.* **70**, 881 (1948).
(19) Kekulé, A., *Ann. Chem. Pharm.* **106**, 129 (1858). Translated in "Classics in the Theory of Chemical Combination," O. T. Benfey, ed., p. 109, Dover Publications, New York, 1963.
(20) Kekulé, A., *Bull. Acad. Roy. Belg.* [2], **19**, 551 (1865).
(21) Kekulé, A., *Ann.* **137**, 129 (1866).
(22) Kekulé, F. A., *Z. Chem.* **3**, 217 (1867).
(23) Kepler, J., "Mysterium Cosmographicum," Tübingen, 1596.
(24) Koestler, A., "The Sleepwalkers," p. 249, Macmillan, New York, 1949.
(25) Kolbe, H., *J. Prakt. Chem.* **2**, 473 (1877). Translated in "Advanced Organic Chemistry," G. W. Wheland, 2nd ed., p. 132, Wiley and Sons, New York, 1949.
(26) Laurent, A., "Méthode de Chimie," p. 408, Paris, 1854. Translated as "Chemical Method," by William Odling, p. 337, London, 1855, reprinted in Benfey, O. T., ed., *op. cit.*, p. 115.
(27) Lipscomb, W. N., *J. Chem. Phys.* **22**, 985 (1954).
(28) Longuet-Higgins, H. C., Roberts, M., *Proc. Roy. Soc. London* **A230**, 110 (1955).
(29) Lucht, M., *J. Am. Chem. Soc.* **73**, 2373 (1951).
(30) McCarty, L. V., Kasper, J. S., Horn, F. H., Decker, B. F., Newkirk, A. E., *J. Am. Chem. Soc.* **80**, 2592 (1958).
(31) Needham, J., "Science and Civilization in China," Vol. II, p. 245, Cambridge University Press, Cambridge, 1962.
(32) *Ibid.*, Vol. I, p. 23.
(33) Pasteur, L., "Lecons de Chimie Professées en 1860," Chemical Society, Paris, 1861. Translated in *Am. J. Pharm.* **34**, 15 (1862).
(34) Pauling, L., Hayward, R., "The Architecture of Molecules," W. H. Freeman, San Francisco, 1964.
(35) Pauling, L., in "Hydrogen Bonding," D. Hadzi, ed., p. 3, Pergamon Press, London, 1958.
(36) Pitochelli, A. R., Hawthorne, M. F., *J. Am. Chem. Soc.* **82**, 3228 (1960).
(37) Plato, "Timaeus," translated by F. M. Cornford, p. 55, Bobbs Merrill Co., Indianapolis, 1959.

(38) Read, J., "Humor and Humanism in Chemistry," p. 87, G. Bell, London, 1947.
(39) Sarton, G., "Introduction to the History of Science," Vol. I, p. 113, Carnegie Institution, Washington, D. C., 1927.
(40) Sementsov, A., ADVAN. CHEM. SER. 61, 72 (1966).
(41) Silbiger, G., Bauer, S. H., J. Am. Chem. Soc. 70, 115 (1948).
(42) Thorndike, L., "History of Magic and Experimental Science," Vol. 6, p. 110, Columbia University Press, New York, 1951.
(43) Toulmin, S., Goodfield, J., "The Architecture of Matter," p. 80, Harper and Row, New York, 1962.
(44) Weizsäcker, C. F., "The Relevance of Science," p. 69 ff., Harper and Row, New York, 1964.
(45) Wislicenus, J. A., Ber. 2, 620 (1869).
(46) Wollaston, W. H., Phil. Trans. 98, 96 (1808). Reprinted in "Foundations of the Atomic Theory," Alembic Club Reprint No. 2, p. 39, E. and S. Livingston, Ltd., Edinburgh, 1948.
(47) Wunderlich, J. A., Lipscomb, W. N., J. Chem. Phys. 22, 989 (1954).
(48) Zhdanov, G. S., Sevast'yanov, N. G., Compt. Rend. Acad. Sci. U.R.S.S. 32, 432 (1941) (in English).

RECEIVED December 29, 1965.

Dreams and Visions in a Century of Chemistry

EDUARD FARBER

American University, Washington, D. C.

In addition to accidental observations, analogies, and inferences by close reasoning, dreams and visions had an important part in the progress of chemistry. Four classes of progress can be distinguished: (1) symbolization and construction of models: Kekulé, van't Hoff, J. J. Thompson; (2) extrapolation in quantity: Wöhler, Sabatier, Kurnakow; (3) projection in time: Kuhlmann, Le Bon, Aston; (4) generalizations: Clausius, Le Chatelier, Ostwald. This list is incomplete and leaves out the failures. Not all those with dreams and visions were as careful as Kekulé was to check and test before publishing. The courage to persist must be combined with a critical evaluation of the facts, and this is especially necessary when solutions are achieved primarily in broad jumps rather than small steps.

We need Kekulé's testimony today as a powerful reminder that chemistry advances not by experiments alone but by a process in which dreams and visions can play an important role. Chemists seem to be particularly inclined to disparage anything that is not experiment; perhaps they still have a guilt complex about alchemy and the speculative periods of the 17th and 18th centuries. In an attitude of defense against speculation, J. C. Poggendorff refused to publish Robert Mayer's paper about "forces in inanimate nature" (1842). This defensive position was fortified by scientific standards of verification, but it also contained an element of prejudice that has been harmful. Results of experimental work were rejected when they would have required a change in cherished assumptions. A prominent example was the measurement by Hermann Helmholtz that the propagation of impulses in nerves takes time and is not, as generally believed, instantaneous (1850).

Kekulé took his dreamlike vision seriously enough to work on it for years, but it had no place in his scientific paper of 1865; he described it at the celebration of this paper 25 years later. While many dreams and visions may never have found their way into the scientific literature, others were mentioned in biographies and letters; all of them should be recognized for their significance in the history and methodology of science. Usually, accounts of such visions delight us as anecdotes, as individual acts that, while too complex for scientific analysis, yet are significant features of scientific progress. Because of this apparent ambiguity, they are not dignified enough to be included in the philosophy of science. For example, Philipp Frank writes (5):

Thus the work of the scientist consists of three parts: (1) setting up principles; (2) making logical conclusions from these principles in order to derive observable facts about them; (3) experimental checking of these observable facts.

Frank here omits the fact that the "principles" are often replaced by symbols or models, and all of them have their particular origins, sometimes in dreams and visions. Yet, he certainly was familiar with a report like that by Sir William Rowan Hamilton (7):

The Quaternions started into life, or light, full grown, on the 16th of October, 1843, as I was walking with Lady Hamilton to Dublin, and came up to Brougham Bridge, which my boys have since called Quaternion Bridge, that is to say, I then and there felt the galvanic circuit of thought close, and the sparks which fell from it were the fundamental equations between i, j, k, exactly such as I have used them ever since.

Frank also knew how Henri Poincaré solved the problem of Fuchsian functions during a sleepless night (25):

I felt them knocking against each other . . . . until two of them hung together, as it were, and formed a stable combination. In the morning, I had established the existence of a class of Fuchs-functions. The results were set down in a few hours.

Occurrences like these two selected from the history of mathematics are frequent enough in the history of science to be accepted as influential, and if they cannot be "explained," they can at least be classified according to distinguishing features. I propose to do just that for 12 instances from the last 100 years in chemistry.

Class I: Symbols to Models

What Kekulé saw in his dream was a snake which "seized hold of its own tail." The picture of a snake in this position was familiar to historians; it was a symbolic reality for alchemists who called it Ouroboros. Had Kekulé seen it in one of the later reproductions? It appears on the title page of a chemical book published in 1690 (Figure 1), on a portrait of Thomas Wright painted in 1740 (24). Injecting the concept

Figure 1. Title page of Andr. Stisser's "Acta Laboratorii Chemici,"
Helmstadt, 1690, with the Ouroboros in the lower left corner.
(Courtesy of Jacob Zeitlin, Zeitlin & Ver Brugge, Los Angeles,
Calif.)

of the tetravalent carbon atom, Kekulé turned the ancient symbol into
a modern model.

Kekulé had to interpret what he saw in his dream; Poincaré had
to take a similar step to understand the symbolic language of the dream.
Jacobus Henricus van't Hoff's vision was much more direct; it showed
him the tetrahedron as the model for the carbon atom in organic compounds. The complete vision occurred to him, as it did on an entirely
different subject to Hamilton, while he was taking a walk. The tetrahedron was not a new model in notions of substance, and Kekulé himself

had come close to it seven years before van't Hoff in a publication (*14*) of 1867:

The four units of affinity of the carbon atom, instead of being placed in one plane, radiate from the sphere representing the atom . . . so that they end in the faces of a tetrahedron . . .

Thus, the extension of formulas into space, from two to three dimensions, had already been seen as desirable before van't Hoff proved that it was necessary and solved the problem of the isomeric lactic acids. However, it was a new approach to use the model to represent asymmetry in organic compounds. What Pasteur had explained as derived from the general asymmetry of the universe was now reduced to a structure of the molecule. The vision was elaborated into a theory, but Hermann Kolbe would perhaps not have attacked it so violently if he had felt that it was only a theory.

Models representing the structure of the atom began to be developed in 1840. The somewhat childish drawings of Dalton and their elaboration by Marc Antoine Augustin Gaudin in 1833 were far exceeded in the new visions of the atom as a vortex by Rankine, Helmholtz, William Thomson (Lord Kelvin), and J. J. Thomson (*11*) or as a planetary system with a central sun. We may question whether these models really were the results of visions—were they not merely analogies? The answer is that all analogies, especially when they combine greatly different things, contain something visionary. The planetary model, in particular, is related to the ancient vision of an analogy between microcosm and macrocosm.

The foregoing three dreams and visions have one aspect in common —they led to the construction or design of models. Kekulé considers his model symbolic; the others can be called ikonic, accepting the definitions given by Frey (*6*), although after making the distinction Frey also introduces the "ikonic-symbolic" model for "a symbolic model for which there is also a completely representational (vollständig abbildendes) ikonic model." In the engineering sciences, a model is most often a small object reproducing essential features of a larger one—e.g., the model of a ship or of a distillation column. In chemistry (and physics) models are enlargements—idealizations that enable us to visualize and to design experiments.

Class II: Extrapolations

Like analogies, extrapolations of a wide range belong to this story also. The first example for a visionary extrapolation comes from a letter (*10*) which Friedrich Wöhler wrote to his friend Liebig on June 25, 1863:

I live completely in the laboratory, busy with the new silicium compound which is generated from silicium-calcium and is deep orange-yellow when pure. I become more and more convinced that it [the yellow compound] is composed in the manner of organic compounds in which carbon is replaced by silicium. Its entire behavior is analogous. In the dark, even in water, it remains quite unchanged; in sunlight, however, it develops, in a kind of fermentation, hydrogen gas and turns snow-white . . . On dry distillation the yellow silicone, as I shall call it, behaves like an organic compound. One obtains hydrogen gas, silicium-hydride gas, brown amorphous silicium (corresponding to coal) and silicium dioxide (corresponding to carbon dioxide).

In my translation I have used the German form of the name for the element silicon, particularly because silicium is more distinctly different from the new word silicon(e), coined by Wöhler.

In contrast to Wöhler's vision, note the conclusion Frederick Stanley Kipping (1863–1949) reached after intensive work on silicones (*15*):

Most if not all the known types of organic derivatives of silicon base have now been considered, and it may be seen how few they are in comparison with those which are entirely organic; as moreover the few which are known are very limited in their reactions, the prospect of any immediate and important advance in this section of organic chemistry does not seem to be very hopeful.

He stated this in 1936. Five years later, important advances began at General Electric Co. and the Dow Chemical Co.

Nikolai Semonowitch Kurnakow (1860–1941) introduced the distinction between Daltonides and Berthollides. He extrapolated from studies of the thallium-bismuth system in combination with (a) the concept of phase according to Gibbs, (b) thoughts about the definite proportions in chemical compounds, published by Franz Wald and supported by Wilhelm Ostwald, (c) the work on variable compositions of minerals by Friedrich Rinne, and (d) other information concerning relationships between properties and composition. Kurnakow designated as "Dalton points" the singular points on diagrams for the relationship of composition to properties (melting points, electric conductivities, etc.): "the composition that corresponds to this point remains constant when the factors of equilibrium change." Besides the Daltonides there is a class of variable compounds comprising the large number of Berthollides (*13, 18*): "Before our eyes, a new and unexplored field opens up, attracting the scientist by its freshness, and promising him rich yields."

In his work on hydrogenations and dehydrogenations with metals as catalysts, Paul Sabatier (1854–1941) encountered many difficulties. The catalysts would sometimes refuse to act. It took time and persistence before the disappointments were explained by a poisoning of the catalysts, especially by sulfur or arsenic. One great idea kept him going (27):

This idea of an intermediary, unstable combination [between catalyst and reagent] has been the beacon that directed all my research on catalysis. Its light will perhaps be extinguished in the future when more powerful new brightnesses will unexpectedly arise; nonetheless, what the beacon has shown will remain as established facts.

In his book on catalysis in organic chemistry Sabatier repeated these words; they were not just a rhetorical embellishment in a speech at a solemn meeting.

Hermann von Helmholtz (1821–1894) described a similar personal experience when he reminisced (9) about his work on an "eye-mirror" at the celebration of his 70th birthday:

Without the assured theoretical conviction that it should be feasible, I would perhaps not have persisted. Since I was in the uncomfortable position quite frequently of having to wait for helpful intuitions, I have gained some experience when and how they arrive which may perhaps become useful for others. Often enough, such intuitions sneak in gently into the thinking, and their significance is not recognized at the start; later on it may be only an accidental circumstance that serves to realize when and under which conditions they have come; otherwise, they are just there and we do not know whence. In other situations they step in suddenly, without effort, as if by association. As far as my experience goes, they never come to the tired brain and not at the writing desk. I always had to turn my problem around and around so much that I had a survey of all its contortions and complications in my head and was able to follow them freely without any writing.

What Sabatier calls idea and Helmholtz calls intuition is not very different from vision, particularly when this is based on extrapolation.

Class III: Predictions

Predictions made from an established system of facts and for a near future are a normal part of science. A visionary element enters when predictions are ventured from a small factual basis and for a distant future. Three examples follow.

Frédéric Kuhlmann (1803–1881) concluded from a long study of ammonia and nitric acid (17, 22):

The ease with which I have been able to transform ammonia into nitric acid indicates that Europe will some day be placed in the condition of the greatest inadequacies from overseas relations for its supply of nitrates; and if the calamities of war were to place us again under the conditions of a blockade, France would get along without India or Peru for assurance of war munitions, for France would always possess animal matter and manganese dioxide.

The vision of a Europe inadequately supplied with nitrates from overseas and the prediction of a France under blockade and having to produce nitric acid for ammunition by oxidizing ammonia came true following the events of 1914, although neither animals nor MnO_2 were used in the process.

In his book on the evolution of matter, Gustave Le Bon (1841–1931) visualized far-reaching consequences; among them (*4, 19*):

Force and matter are two different forms of one and the same thing. The power to dissociate matter freely would put at our disposal an infinite source of energy and would render unnecessary the extraction of that coal the provision whereof is rapidly becoming exhausted.

At the end of his Nobel Prize lecture in 1922, Francis William Aston (1877–1945) explained that 1 gram of hydrogen would release the equivalent of 200,000 kw. hours if it were completely converted into helium and then added this vision of what could happen (*2*):

Should the research worker of the future discover some means of releasing this energy in a form which could be employed, the human race will have at its command powers beyond the dream of scientific fiction . . .

If, however, the reaction should get out of control, it would be "published at large to the universe as a new star."

Class IV: Universal Generalizations

In the year in which Kekulé published his benzene formula Rudolf Clausius expressed the basic laws of the universe which correspond respectively to the two laws of the mechanical theory of heat in the following simple form (*3, 16*):

(1) the energy of the universe is constant,

(2) the entropy of the universe aspires to a maxium.

Clausius uses the verb "strebt . .zu," which is more adequately rendered by "aspires" than by the usual translation "tend toward." He coined the word entropy for the quantity:

$$S = S_o + \int \frac{dQ}{T}$$

Previously, he had called "verwandelt" the heat transferred from high to low temperature in the Carnot cycle. I assume he selected the letter S because the letters on both sides of it, Q, R, T, U, and V already had found their definite meanings in thermodynamics by a generally accepted convention. Entropy stands for the "transformation content" (Verwandlungsinhalt), and its maximum meant a minimum in the possibility of further "transformation."

With this universal generalization Clausius went beyond anything that can be called theory and even beyond the generalization William Thomson (1822–1907) had reached in 1852: "there is at present in the material world a universal tendency to the dissipation of mechanical energy."

Josiah Willard Gibbs used the two Clausius sentences as a motto and placed them in front of his lengthy paper "On the equilibrium of heterogeneous substances."

The vision of a general tendency recurs in the law which Henri Louis Le Chatelier published (20) in 1884:

Every system in stable chemical equilibrium, submitted to the influence of an exterior force which tends to cause variation either of its temperature or its condensation . . ., can undergo only those interior modifications which, if they occurred alone, would produce a change of temperature, or of condensation, of a sign contrary to that resulting from the exterior force.

This law of an interior counteraction against an exterior force was of great help to Walther Nernst and to Fritz Haber when they designed the optimum conditions for combining nitrogen with hydrogen in synthesizing ammonia.

In the spring of 1890, Wilhelm Ostwald (1853–1932) left Leipzig to persuade a friend in Berlin to write a textbook of physics from the standpoint of energetics. A long discussion, joined by others, extended far into the night (23):

I . . . slept for a few hours, then suddenly awoke immersed in the same thought and could not go back to sleep. In the earliest morning hours I went from the hotel to the Tiergarten, and there, in the sunshine of a glorious spring morning, I experienced a real Pentecoste, an outpouring of the spirit over me. . . . This was the actual birth-hour of energetics. What a year before, at that first, sudden sensation in my brain that was the conception of the thought, had confronted me as rather strange, even with the taint of frightening newness, now it proved to belong to myself, so much so that it was a life-supporting part of my being. . . . at once everything was there, and my glance only had to glide from one place to the other in order to grasp the whole new creation in its perfection.

When Ostwald wrote this, he was unaware of what Helmholtz had said about his "intuitions" almost 40 years before. Ostwald had gone through other experiences of "lightning-like" visions, but this was the most important because it was the most universal generalization he could reach.

Though the thought was completely subjective—"a life-important part of my being"—it embraced an objective totality. The apparent paradox recurs in all dreams and visions; it is especially great in the universal generalizations.

Conclusion

Many more examples could have been cited, extending from recent chemistry to dreams and visions at other times, on different subjects, starting with Johannes Kepler's dream of a trip to the moon (26), mentioning René Descartes, continuing with Emanuel Swedenborg (21, 28), and ending with Arnold Toynbee (29). The classification could have been based on the kind of visualized picture or on the scientific consequences, and the psychological conditions would have formed the best basis if only the biographical information were available in sufficient depth. The classification I used has more the function of an aid to the memory than the character of a deeply unifying system. In addition, there is a certain increase in range in the progression from the first class to the fourth.

The main conclusion is that dreams and visions deserve to be recognized, without ridicule or pretense, as having an important place, even in modern chemistry. They must be treated with critical respect. Danger looms as much in their presence as in their absence. Imagine what might have happened if Otto Hahn and his group had not stuck to the "false trans-uranium elements" which they later found to be a scientific error: "A completely unexpected reaction, forbidden by physics, the break-up of the highly charged element uranium into barium and . . . . krypton" (8) opened a new era to those who had the vision, though they were not the first to carry out the experiment.

In describing the "how and when," Helmholtz tried to arrive at a prescription for inviting "intuitions." These prescriptions, together with the experiences cited, lead to a further conclusion—they required a relaxed patience that nevertheless was charged with the vital tension necessary to solve a problem, followed by a critical objectivity towards the vision. This is entirely different from a "crash program" that diverts mental functions from the problem to pretentious and premature communications.

Dreams and visions are necessary, but they are not sufficient. Drawing attention to the rightful place of a component should never imply that it is more than a part, and certainly it is not identical with the whole. Great advances in chemistry have come through improved experimental skill and accuracy. Increased reliability of analysis led to the discovery of lithium and of the first three "nationalistic" elements: scandium, gallium, and germanium. Greater precision in measuring optical emission spectra was instrumental in finding several new elements and atomic structures. The complementary nature that exists between visions and experimental skills does not mean that they must be distributed over different personalities; the same man can use one or the other at different times in his work.

Appendix

Kekulé and the Serpent. In the book resulting from the Kekulé Symposium in London, 1958, P. E. Verkade writes (30):

On the other hand, it is to be noted that the snake biting its own tail had also played a part early in Kekulé's life. In 1847 he appeared as a witness in a trial for the murder of Countess Görlitz, who lived next door to his father at Darmstadt; this murder was coupled with a theft of jewelry, including a ring that consisted of two intertwined metal snakes biting their own tails. The incident in question made a deep impression on Kekulé and may have led to the famous dream.

In this account the facts are inaccurate, and the conclusions are no less arbitrary than my conjecture of a connection with the ancient symbol of Ouroboros. In 1847 Kekulé saw the fire in his neighbor's house, and he was therefore called in as an eye witness in the murder trial, which took place in Darmstadt, March 11 to April 11, 1850. The expert witness at the trial was Justus Liebig. Richard Anschütz describes the events from the sources available to him and adds (1):

Bei der Erzählung der Aufstellung der Benzoltheorie werden wir Kekulé das Bild von der Schlange, die sich in den Schwanz beisst, auf die Kohlenstoff-Kette anwenden sehen, die sich zum Ringe schliesst.

There is no hint here that Kekulé saw the ring or that he knew about its form. Those who are sensitive to form and logic in historiography will notice with distress that Anschütz reverses the relationships between "the carbon chain . . . which closes in on itself to a ring . . ." and "the picture of the snake which bites its own tail."

After this incidental remark I return to our problem. Finger rings with various forms of special emphasis on the "return into itself" were not unusual; several of them are shown in a relatively recent book on jewelry design (12). If Kekulé had continued to have "a deep impression" of any of these rings, including that of the Countess, would he not have mentioned it in the story he leisurely told about his dream? The snake and its magic position did not have any function after the dream. Could not the picture of any ring, in fact of any circle, have served to initiate his thoughts about the benzene "ring"? We could imagine that it might have happened this way, but the historical reality was different. The action Kekulé saw in his dream showed him what to do.

Literature Cited

(1) Anschütz, R., "August Kekulé," Vol. 1, p. 19, Verlag Chemie, Berlin, 1925.
(2) Aston, F. W., "Les Prix Nobel en 1922," cited in Farber, E., "Nobel Prize Winners in Chemistry," p. 90, Abelard-Schuman, New York, 1963.
(3) Clausius, R., *Pogg. Ann.* **125**, 353 (1865).
(4) Farber, E., *Chymia* **9**, 198 (1964).
(5) Frank, P., "Philosophy of Science, the Link Between Philosophy and Science," p. 43, Prentice-Hall, Englewood Cliffs, 1957.

(6) Frey, G., *Proc. Colloq. Div. Phil. Sci.*, Dordrecht, **1961**, 96.
(7) Graves, R. P., "Life of Hamilton," Vol. 2, p. 434, Dublin, 1882.
(8) Hahn, O., *Naturw. Rundschau* **15**, 43 (1962); **18**, 90 (1965).
(9) Helmholtz, H., "Vorträge und Reden," 5th ed., Vol. 1, pp. 12, 15, Vieweg, Braunschweig, 1902.
(10) Hofmann, A. W., "Zur Erinnerung an vorangegangene Freunde," Vol. 2, p. 99, Vieweg, Braunschweig, 1888.
(11) Ihde, A. J., "The Development of Modern Chemistry," p. 475, Harper & Row, New York, 1964.
(12) Jossic, Y. F., "1050 Jewelry Designs," plate 17-2 (rings from Germany, 16th-18th century), A. A. Lampe, Philadelphia, 1946.
(13) Kauffmann, G. B., Beck, A., *J. Chem. Ed.* **39**, 44 (1962).
(14) Kekulé, F. A., *Z. Chem.* **3**, 217 (1867).
(15) Kipping, F. S., *Proc. Roy. Soc.* **A159**, 147 (1936).
(16) Koenig, F. C., "Men and Moments in the History of Science," ed. E. M. Evans, p. 57, University of Washington Press, Seattle, 1959.
(17) Kuhlmann, F., *Ann.* **20**, 223 (1847).
(18) Kurnakow, N. S., *Z. Anorg. Chem.* **88**, 109 (1914).
(19) Le Bon, G., "The Evolution of Matter," translated by F. Legge, 2nd ed., pp. 8, 51, Walter Scott Publishing Co., London and New York, 1907.
(20) Le Chatelier, H. L., *Compt. Rend.* **99**, 786 (1884); translated in Leicester, H. M., Klickstein, H. S., "A Source Book in Chemistry," p. 481, McGraw-Hill, New York, 1952.
(21) Meyer-Lune, I., "Swedenborg, eine Studie über seine Entwicklung," Leipzig, 1922.
(22) Mittasch, A., "Salpetersäure aus Ammoniak," p. 15, Verlag Chemie, Weinheim, 1953.
(23) Ostwald, W., "Lebenslinien, eine Selbstbiographie," Vol. 2, p. 160, Velhagen & Klasing, Berlin, 1933.
(24) Paneth, F., *Durham Univ. J.* **2**, 111 (1941), reproduced in "Chemistry and Beyond," Dingle, H., Martin, G. R., eds., p. 95, Interscience, New York, 1965.
(25) Poincaré, H., "Science and Method," p. 52, Dover, New York, 1913.
(26) Rosen, E., *Proc. 10th Intern. Congr. History Sci.*, Ithaca, **1962**, 81.
(27) Sabatier, P., *Ber.* **44**, 2001 (1911).
(28) Stroh, A. H., Ekelof, G., *Isis* **23**, 459, 520 (1935).
(29) Toynbee, A. J., "A Study of History," Vol. 10, p. 139, Oxford University Press, Oxford, 1954.
(30) Verkade, P. E., *Theoret. Org. Chem., Papers Kekulé Symposium, London*, **1958**, xvi, 1959.

RECEIVED September 24, 1965.

9

The Development of Strain Theory

AARON J. IHDE

University of Wisconsin, Madison, Wis.

To account for the apparent instability of rings with less than five carbon atoms, Baeyer suggested in 1885 a strain of the valencies of carbon away from the normal tetrahedral angle of 109° 28′. The concept proved valuable in interpreting structural problems connected with small rings, bridged rings, and structures found in natural products such as sterols and terpenes. Application to six-membered and larger rings caused problems until it was realized that such rings are strainless owing to their ability to take on a puckered conformation. Heats of combustion, dipole moments, spectra, as well as chemical evidence have generally been in accord with strain predictions based on examination of models.

It was commonly believed prior to 1880 that organic compounds belonged to only two classes—open chain and aromatic. The ring structure of benzene, the parent compound of the aromatic class, had become generally accepted, and many derivatives of benzene were known. Also known were such representatives of the aromatic class as naphthalene, anthracene, phenanthrene, and pyridine—compounds which figured prominently in many syntheses. A limited number of compounds containing five-membered rings had been purified and studied—e.g., phthalic anhydride, isatin, indigo, indole, furan, pyrrole, and several lactones, but the structural characteristics had not yet been established to everyone's satisfaction. It should be noted that the known five-membered rings contained an atom of nitrogen or oxygen. No rings were known which contained carbon atoms exclusively and which had more or less than six carbon atoms.

In 1876 Victor Meyer (37) published a paper in which he considered it highly unlikely that rings of less than six carbon atoms would ever be encountered (Figure 1). He recalled that reactions which should presumably lead to the formation of three-carbon rings had always resulted in producing unsaturated open-chain compounds and pointed out there

Courtesy of the Edgar Fahs Smith Collection

Figure 1.   Victor Meyer

had been less effort toward discovering four- and five-membered rings, but except for one unconfirmed formula for anthracene, there was little evidence to suggest the occurrence of rings smaller than six carbon atoms.

Nearly a decade later Adolf von Baeyer (3), known as a brilliant experimenter who was not prone to theoretical speculation, advanced his Spannungs Theorie (Strain Theory) as a short addendum to a paper dealing with preparation and properties of acetylene derivatives (Figure 2).

Since the di- and triacetylenes described in the body of the paper are highly reactive and even explosively unstable, he felt an obligation to deal with the instability of what he chose to consider a ring system. However, a large part of his research had been with aromatic compounds; hence there was added incentive to ponder the problem of ring stability. Introducing the theory as an apparent afterthought in an experimental paper was characteristic of him since he was far more comfortable as an experimenter than as a theorist.

After observing that a consideration of spatial arrangements can lead to understanding ring closure, Baeyer summarized what was then known about the nature of carbon atoms and added a statement of his own suggesting that the four valences of carbon make an angle of 109° 28′ with one another. He then argued, "The direction of these attractions can undergo a diversion which causes a strain which increases with the

Courtesy of the Edgar Fahs Smith Collection

Figure 2.   Johann Friedrich Wilhelm Adolf von Baeyer

size of the diversion" (3). The diversion of the bonds was compared with the distortion of elastic springs. Clearly, such distortion led to strain and consequent instability. In open-chain compounds strain was avoided since a zig-zag arrangement of carbon atoms was normal, but if the ends of the chain were joined to form a ring, strain was inevitable. Baeyer went on to show that in no case did ring size permit the normal tetrahedral angle of 109° 28′ to exist. The closest approach to this angle was the five-carbon ring where it was 108°. In both larger and smaller rings the angular distortion or strain became greater. He calculated the angular strain for various cycloparaffins, using the equation,

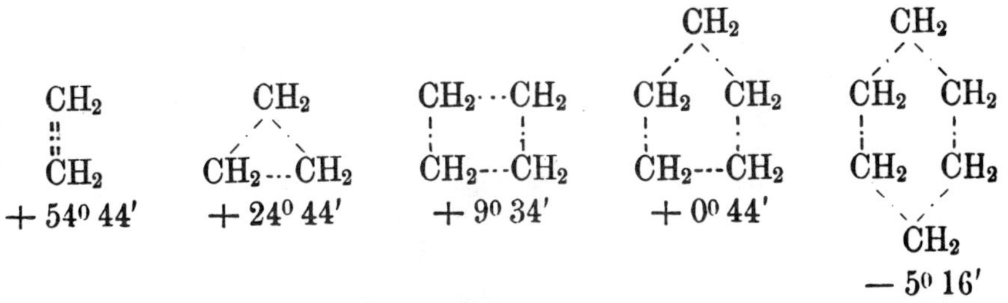

$$CH_2$$
$$\overset{\shortparallel}{C}H_2$$
$$+ 54^0\ 44'$$

$$CH_2$$
$$CH_2 \cdots CH_2$$
$$+ 24^0\ 44'$$

$$CH_2 \cdots CH_2$$
$$CH_2 \cdots CH_2$$
$$+ 9^0\ 34'$$

$$CH_2$$
$$CH_2 \quad CH_2$$
$$CH_2 \cdots CH_2$$
$$+ 0^0\ 44'$$

$$CH_2$$
$$CH_2 \quad CH_2$$
$$CH_2 \quad CH_2$$
$$CH_2$$
$$- 5^0\ 16'$$

Figure 3. Calculated strains in carbon rings containing two to six members from Baeyer's paper (3). Note that the calculated angle for cyclobutane should be + 9° 44'

½ (109° 28' — actual bond angle), to obtain the values shown in Figure 3. As is evident from the equation, a positive value is associated with bond angles less than 109° 28', a negative value with bond angles larger than the normal, as in cyclohexane.

Baeyer pointed out that the ring of "dimethylene" is "broken by hydrogen bromide, bromine, and even iodine" whereas trimethylene is split only by hydrogen bromide. The four- and six-membered rings "are rare and found in complicated compounds." Then he passed off the observation as being of little consequence, commenting that six-membered rings are found almost entirely in the form of hydrogen-deficient benzene; he closed the paper by remarking that he intentionally disregarded thiophene, lactones, and similar compounds because other elements are present therein. He was evidently unfamiliar with cyclohexane which Felix Wreden and B. Znatowicz (74) had prepared in impure form by hydrogenating of benzene.

Synthesis of Small Rings

At the time Baeyer's paper was written, the supposed nonexistence of small rings had just been demolished by one of his own students, William Henry Perkin, Jr., and Baeyer called attention to the greater stability of three- and four-membered rings over double bonds.

When the younger Perkin prepared to study in Germany in 1880, Victor Meyer's paper was one which he translated to give himself familiarity with the language. Two years later, when Perkin was working on the preparation of benzoylacetic ester in Baeyer's laboratory in Munich, he had an opportunity to meet Victor Meyer. The young chemist asked the master if he were still convinced of the impossibility of synthesizing rings smaller than six carbon atoms. Meyer spent an evening with Perkin discussing ring formation and arguing, on the grounds of the nonexistence of three-, four-, and five-membered rings, the unlikelihood of ever preparing such compounds. When Perkin declared that

Figure 4.   William Henry Perkin, Jr. about 1890

he intended to attempt their preparation, Meyer expressed admiration for the young man's enthusiasm but advised him, at this early stage of his career, to turn to something more likely to give positive results. On being informed of the discussion, Baeyer said that he too was skeptical about the likelihood of preparing smaller rings, pointing out that they had never been encountered in nature (*46;* this paper, the First Pedler Lecture, is autobiographical and deals with the early synthesis of ring compounds).

Emil Fischer also expressed skepticism when he visited the Munich laboratory, suggesting that even if suitable methods were employed to obtain the desired ring closure, the compounds would be formed in such limited quantity and possess so little stability that it would be difficult to establish their existence.

Despite his initial skepticism, Baeyer frequently brought up the problem of small rings and gave Perkin the impression that he would not look upon such studies with disfavor. The studies on benzoylacetic ester had been disappointing since it was not possible to obtain products in a reasonable state of purity in his day when low pressure distillation was still to be developed. In one of his experiments, where he condensed propyl bromide with acetoacetic ester, the thought occurred to Perkin that possibly propylene dibromide might condense with sodium acetoacetate to give acetyltetramethylene carboxylic ester with a four-membered ring (*46*).

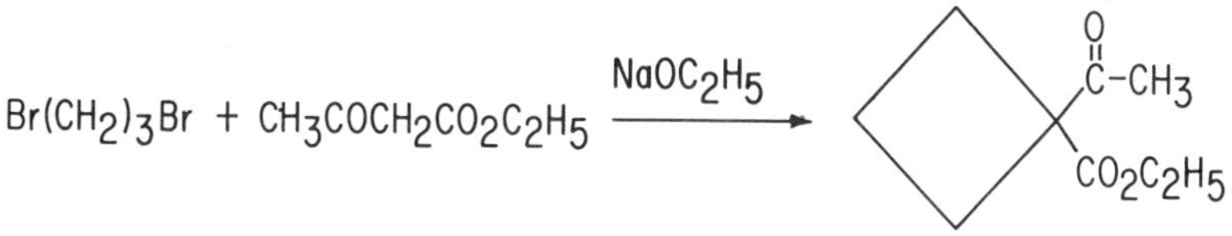

Br(CH2)3Br + CH3COCH2CO2C2H5
NaOC2H5
C-CH3
CO2C2H5

After condensing trimethylene bromide with the sodium derivative of acetoacetic ester, Perkin obtained a product which yielded the elementary analysis for the expected compound. Hydrolysis yielded a crystalline acid, whose analysis agreed closely with that calculated for acetyltetramethylene carboxylic acid. However, the expected decomposition, with the elimination of carbon dioxide or an acetyl group, did not occur although this would be expected of a substituted acetoacetic ester.

Baeyer considered the synthesis worthy of immediate communication to the Bavarian Academy, and it was subsequently published in *Berichte* (*40*). Three years later Perkin learned that the reaction had not taken the reported direction and that no four-carbon ring had been formed after all. Attempted recrystallization of the acid from boiling water was accompanied by evolution of carbon dioxide and failure of the acid to separate on cooling. The soluble product proved to be, not acetyltetramethylene carboxylic acid, but acetylbutyl alcohol (*44*). Perkin now postulated the formation of an oxygen-containing ring com-

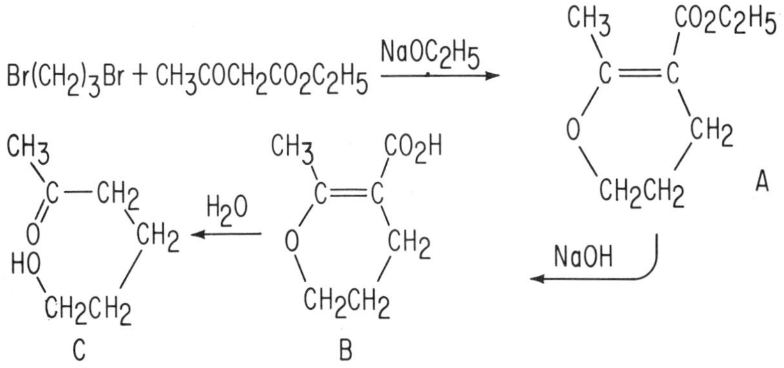

Br(CH2)3Br + CH3COCH2CO2C2H5
NaOC2H5
CH3
CO2C2H5
CH2
CH2CH2
A
NaOH
CO2H
H2O
B
HO
C

pound (A) instead of the compound he had considered to contain a four-membered ring and which had given the apparently correct elementary analysis for $C_9H_{15}O_3$. Saponification then yields the corresponding acid (B), and subsequent boiling in water caused decarboxylation and ring opening to yield acetylbutyl alcohol (C)—a compound first prepared and unambiguously identified at Munich by Lipp (35) a year earlier.

Before the error was discovered, however, Perkin had exploited the new approach and condensed trimethylene bromide with the sodium derivative of malonic ester (41). He reasoned that the product should be either tetramethylenedicarboxylic ester or allylmalonic ester.

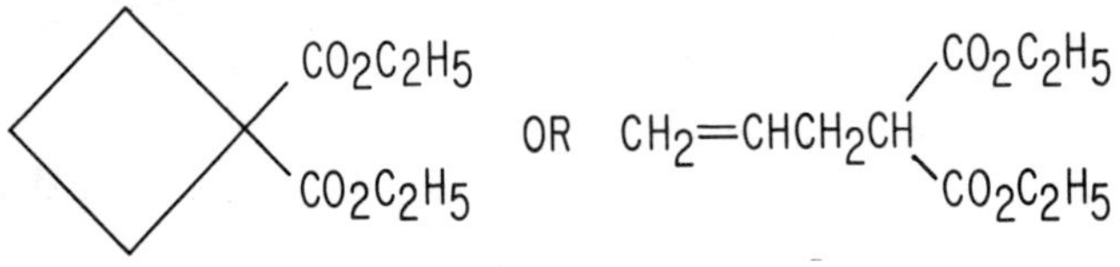

On decarboxylation, the corresponding acid would be formed.

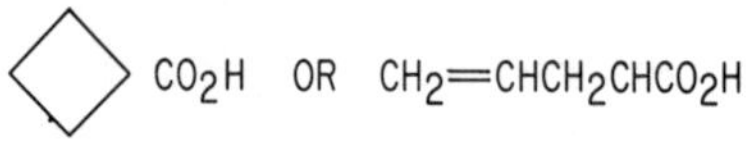

Careful study of the acid, utilizing methods like magnetic rotation and refractive index whereby physical properties might be correlated with chemical constitution, eliminated allylacetic acid as a possibility and suggested the successful synthesis of a four-membered ring.

Slightly earlier, however, evidence for synthesis of such a ring had been published by Markovnikov and Krestovnikov (36). They heated ethyl α-chloropropionate with dry sodium ethylate to obtain an ester which hydrolyzed to a crystalline acid which was named tetrylenedicarboxylic acid.

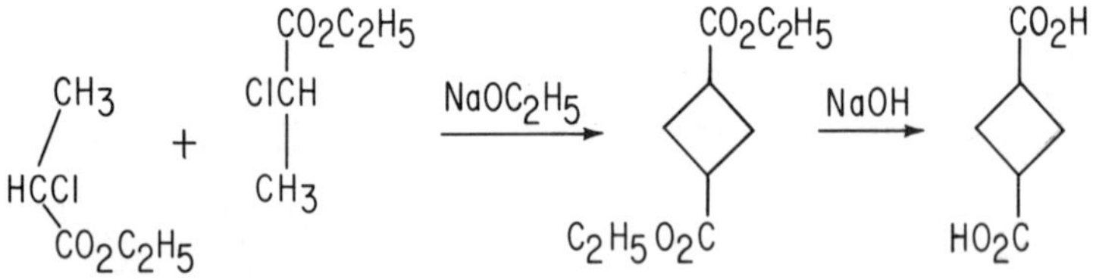

Perkin called attention to the preparation in his 1883 paper, and nearly two decades later he and Haworth verified the synthesis (47). Over the years the work of Markovnikov and Krestovnikov was accepted as the earliest synthesis of a cyclobutane ring. The acid was resynthesized in several laboratories (13, 26, 70), but its structure was never questioned even though discussions of the reaction mechanism raised problems. Finally, Deutsch and Buchman (*19*) recognized the acid to be identical with the 1-methyl-1,2-cyclopropane dicarboxylic acid first prepared and correctly identified in 1924 by Staudinger *et al.* (*65*) and later prepared and studied in other laboratories. Thus, the work of Deutsch and Buchman revealed that Perkin was the first to synthesize a compound with a cyclobutane ring whereas Markovnikov and Krestovnikov created a cyclopropane derivative in 1881. Deutsch and Buchman (*19*) were successful in preparing the 1,3-cyclobutane dicarboxylic acid erroneously postulated by Markovnikov and Krestovnikov and never prepared before 1950.

Perkin undertook the preparation of the cyclopropane analog of his presumed cyclobutane derivative in 1884, treating ethylene bromide with the sodium compound of malonic ester and obtaining a compound that appeared to be what he sought (*42*). Soon thereafter, his interpretation was challenged by Fittig and Roeder, who in their studies on lactone formation had prepared the same compound and suggested it to be vinylacetic acid (*22, 23*). Perkin ultimately established the correct-

$$\triangleright\text{CHCH}_2\text{COOH} \qquad\qquad \text{H}_2\text{C}=\text{CHCH}_2\text{COOH}$$

PERKIN   FITTIG

Apparently Perkin and Baeyer were unaware that cyclopropane itself had been prepared in 1882 by Freund (*24*), who treated propylene bromide with sodium. The product was very impure, however, and it was not until 1907 that pure cyclopropane was prepared by Willstätter (*72*).

Perkin also sought to prepare a five-membered ring. A synthesis along the lines which had been successful for three- and four-membered rings could not be used until 1894 because tetramethylene bromide was unavailable (*45*). A successful synthesis was accomplished in 1885, however, by taking advantage of the fact that when di- or trimethylene bromide reacts with malonic ester, a side reaction leads to the production of higher boiling substances resulting from the reaction of a molecule of dibromide with two molecules of malonic ester (*43*). Perkin, recog-

$$\text{BrCH}_2\text{CH}_2\text{CH}_2\text{Br} + 2\text{CH}_2(\text{COOEt})_2 \xrightarrow{\text{NaOC}_2\text{H}_5} \begin{matrix} \text{C}_2\text{H}_5\text{O}_2\text{C} \\ \diagdown \\ \text{CHCH}_2\text{CH}_2\text{CH}_2\text{CH} \\ \diagup \\ \text{C}_2\text{H}_5\text{O}_2\text{C} \end{matrix} \begin{matrix} \text{CO}_2\text{C}_2\text{H}_5 \\ \diagup \\ \\ \diagdown \\ \text{CO}_2\text{C}_2\text{H}_5 \end{matrix}$$

nizing that the compound would still form a sodium derivative, treated this with bromine to obtain a pentamethylenetetracarboxylic ester (or on hydrolysis, the acid).

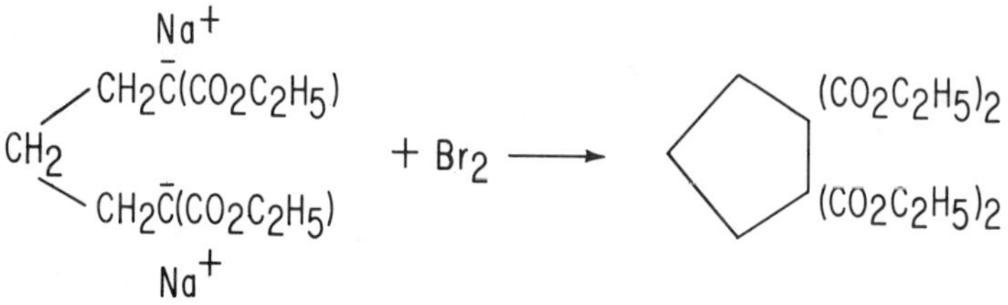

$$
\begin{array}{c}
Na^+ \\
CH_2\overset{-}{C}(CO_2C_2H_5) \\
CH_2 \\
CH_2\overset{-}{C}(CO_2C_2H_5) \\
Na^+
\end{array}
\quad + \ Br_2 \ \longrightarrow \quad
\begin{array}{c}
(CO_2C_2H_5)_2 \\
(CO_2C_2H_5)_2
\end{array}
$$

Obviously by 1885 there was enough knowledge about the preparation and properties of small rings that Baeyer felt confident in advancing his strain theory. The theory was actually published before a five-membered ring had been synthesized. Perkin reports (46) that Baeyer was deeply interested in this compound since its stability confirmed his speculations about five-membered rings.

Thermochemical Aspects of Strain

In the last paragraph of his paper Baeyer (3) suggested that thermochemical measurements should further elucidate strain considerations. When heat of combustion values became available for a series of alicyclic compounds, a correlation with ring strain was possible. Table I gives values for heats of combustion of some of the cycloparaffins, expressed in kilocalories per CH_2 group.

Table I. Heats of Combustion of Some Cycloparaffins

CH_2 groups in molecule	Heat of Combustion per CH_2 group in kcal.	Reference
3	166.6	(34)
4	164.0	(33)
5	158.7	(33, 69)
6	157.4	(33, 69)
7	158.3	(33, 69)
8	158.3	(33, 69)
9	158.9	(33, 69)
10	158.6	(33, 69)
15	157.1	(52)
17	157.0	(52)
30	156	(52)
Aliphatic compounds ca. 157		

The heat of combustion for a single CH_2 group in aliphatic compounds is 157–159 kcal. Thus, cycloparaffins with five or more carbon atoms

show no significant deviation from methylene groups in aliphatic compounds. In the case of cyclobutane and cyclopropane the higher values
are consistent with the greater energy expected in the compounds because of the additional energy required to close a three- or four-membered ring where strain is involved.

It will be observed that the heat of combustion for rings larger
than five does not rise above the normal value, suggesting an absence
of strain in such rings. The reason for this will be examined in the section
dealing with large rings. Thermochemical measurements have also
proved important in connection with many other compounds where
strain is possible, but discussion will be delayed until such compounds
are examined.

The Problem of Large Rings

Table I shows that the heat of combustion per methylene group
remains essentially constant for all compounds larger than cyclobutane.
Such behavior would be expected for cyclopentane since the ring is
almost strainless. In the case of cyclohexane, however, Baeyer calculated a strain of $-5°$ 16', and here bond strain should be reflected in an
increased heat of combustion per methylene group. Strain calculations
for larger rings would give even larger negative values—i.e., $C_7H_{14} =
-9°$ 33', $C_8H_{16} = -12°$ 46', $C_{15}H_{30} = -23°$ 16', $C_{20}H_{40} = -26°$ 16',
assuming that all carbon atoms lie in the same plane. This assumption
was taken to be implicit in Baeyer's strain theory although he nowhere
extrapolated beyond cyclohexane in his paper. Other organic chemists
$(6, 50, 71)$ proceeded to discuss ring strain as if the carbon atoms lay
in the same plane. Meyer and Jacobson's text (38) even included the
formula, $\alpha = \dfrac{180}{n} -35°$ 16' (where α represents angular strain and n
the number of carbon atoms) for calculating strain in rings of any size.

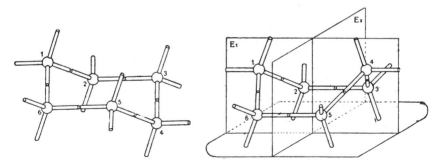

Journal für Praktische Chemie

Figure 5.   Chair and boat models of cyclohexane (39)

The belief in planar rings persisted long after 1890 when H. Sachse (57, 58) published his first paper on strain-free cyclohexane. He pointed out that if a six-membered ring were puckered, two strain-free forms would be possible (later identified as "chair" and "boat" forms) (Figure 5). The suggestion attracted little attention for nearly three decades, in part because isomeric forms of cyclohexane could not be isolated.

In 1918 Ernst Mohr (39) published a significant paper in which, with the aid of drawings of models, he showed the possibility of strain-free rings in various kinds of compounds. Drawing upon the new knowledge of the structure of diamond arising out of the x-ray studies of W. H. and W. L. Bragg, he recognized verification of the tetrahedral model of van't Hoff and saw its fundamental implications in connection with structure of organic ring compounds. He recognized the chair form of cyclohexane as a fragment of the diamond crystal lattice.

In the case of cyclohexane, Mohr argued that the chair and boat forms of the ring should be easily interconvertible; hence isomers could not be isolated successfully. However, he predicted that the compound, decalin, should exist in cis and trans forms which should be separable (Figure 6). Because the two rings would share a common bond, transformation between forms would require bonds to be broken and atoms to be rearranged.

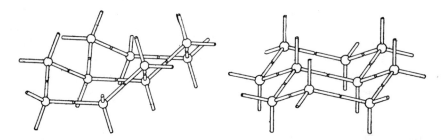

Journal für Praktische Chemie

Figure 6.   Models of cis- *and* trans- *decalin* (39)

The laboratory of Jacob Böeseken (10, 18) in Delft soon produced evidence to support Mohr's views when cyclohexane-1,2-diol was shown to exist in two forms. One form markedly increased the conductivity of a boric acid solution and formed an acetonide; the other did neither. It was concluded that the complex-forming diol was cis, the other trans.

In 1925 Walter Hückel (27, 28) successfully prepared and isolated the isomers of decalin which had been postulated by Mohr. Since the cis and trans forms cannot undergo interconversion without bond breakage, the isomers can be separated by fractional distillation. In subsequent years Hückel contributed extensively to the understanding of structure and stability of ring compounds (*29, 30*).

In still another direction, the supposed instability of multimembered rings was demolished in 1926 through the work of Leopold Ruzicka on perfume ingredients isolated from musk and civet. Analytical and synthetic studies revealed the compounds muscone and civetone to be ketones made up of rings containing 15 and 17 carbon atoms, respectively (*53, 54, 55, 56*).

$$CH_3 \; CH-CH_2$$
$$\bigg| \qquad \diagdown C=O$$
$$(CH_2)_{12} \diagup$$

MUSCONE

$$CH(CH_2)_7$$
$$\| \qquad \diagdown C=O$$
$$CH(CH_2)_7 \diagup$$

CIVETONE

Studies by Ruzicka's laboratory on preparing large rings revealed that such rings are prepared fairly easily, and once formed, the compounds are stable since they exist in a puckered form, thereby setting up no significant strains. Heats of combustion on such compounds are in accord with the concept of strainless rings. Ruzicka's studies included rings containing up to 30 carbon atoms.

Experience has shown that three- and four-membered rings are not produced in high yields. In this case strain is unavoidable. Five-, six-, and seven-membered rings are made easily and are stable. Larger rings are not difficult to make, but yields are poor, particularly in the C_8 to C_{14} range with a minimum at C_{10}. In part, the low yield is related to the low rate of formation. The probability that the two ends of a chain will be in a favorable position for ring closure is poor. Further, the ring-closure reaction competes with intermolecular reactions which yield polymeric molecules. The minimum yield at C_{10} and thereabouts has been ascribed to a crowding of hydrogen atoms leading to H—H repulsions. Finley (*21*) has made a fine historical study of the synthesis of rings, both large and small, so the subject will not be pursued further here.

Reactivity

At the time it was introduced, strain theory was correlated with the natural occurrence of ring compounds and the ease with which rings could be synthesized. It was observed that five- and six-membered rings were easily prepared in good yield in the laboratory and that the latter type of compound was common in nature. Such rings are stable. They were much harder to open than three- and four-membered rings. This also proved true of heterocyclic compounds. Later research, such as measurements of dipole moments of simple compounds of oxygen, sulfur, and nitrogen, permitted calculations of bond angles and revealed values between 100° and 110°. Hence, these atoms create no divergence in strain features from those in rings composed exclusively of carbon atoms.

Behavior of hydroxy acids was being studied actively in Fittig's laboratory at the time the strain theory was introduced. It was evident that the position of the hydroxyl group in relation to the carboxyl group was of crucial importance in connection with the product formed in reactions where water was removed from the molecule. Heating α-hydroxy acids results in intermolecular condensation to form lactides with six-membered rings.

$$
\begin{array}{ccc}
\text{R CHO} \boxed{\text{H} \quad \text{HO}} \text{OC} & & \text{R CH-O-CO} \\
| \quad\quad\quad\quad | & \longrightarrow & | \quad\quad\quad | \quad + 2H_2O \\
\text{CO} \boxed{\text{OH} \quad \text{H}} \text{OCHR} & & \text{OC-O-CHR}
\end{array}
$$

With β-hydroxyacids water is removed to form α, β-unsaturated acids (with traces of β, γ-unsaturation). With γ- and δ-hydroxy acids cyclization occurs with the formation of lactones. When the hydroxyl group

$$
\begin{array}{ccc}
\text{R CHCH}_2\text{CH}_2\text{C=O} & & \text{R CHCH}_2\text{CH}_2\text{CO} \\
| \quad\quad\quad\quad | & \longrightarrow & \boxed{} \quad + H_2O \\
\text{OH} \quad\quad \text{HO} & & \text{O}
\end{array}
$$

is on the epsilon position there is sometimes lactone formation (7-membered ring) but more commonly δ, ϵ or ϵ, ζ unsaturation or formation of a linear polyester. Acids with the hydroxyl group further removed tend to form polyesters or to become unsaturated.

Such behavior is in accord with expectations; γ- and δ-lactones contain strain-free rings. Larger rings would also be strain free but the probability that the reactive centers would come into proximity decreases as the hydroxyl and carboxyl groups become further removed from each other. The α and β positions are highly unfavorable for lactone formation because of the strain involved in small rings.

Of the numerous studies on lactone formation, those of Sebelius (61) are particularly significant. He made extensive studies with γ- and δ-hydroxy acids to ascertain the influence of substituent groups on the equilibrium constant of the reaction and the equilibrium concentration of lactone.

Sebelius' studies showed that at equilibrium, the γ-lactones show a higher concentration than δ-lactones of similar structure, although lactone formation takes place more slowly. However, the γ-lactones also hydrolyze more slowly. The work also revealed that substituent groups, especially on the α and γ (or δ) positions improved the equilibrium concentration of lactone. A few examples are shown in Table II.

Table II

Lactone	Equilibrium concentration in % lactone	Equilibrium constant
γ-Butyrolactone	72.8	2.68
α-Methyl-γ-butyrolactone	95.4	20.5
γ-Methyl-γ-butyrolactone	93.0	13.3
δ-Valerolactone	9.0	0.099
α-Methyl-δ-valerolactone	16.5	0.198
δ-Methyl-δ-valerolactone	21.2	0.269
δ-Dimethyl-δ-valerolactone	25.1	0.335

Ring closure can proceed only when the reactive groups of suitable compounds are brought into proper proximity, a condition seldom expected. Energy conditions must be favorable for bending the chain into such proximity. Once a favorable position is attained, a certain amount of activation energy is essential for closure. If strain is involved, there will be resistance to closure, and even greater energy is required, frequently involving an input of several kilocalories per mole.

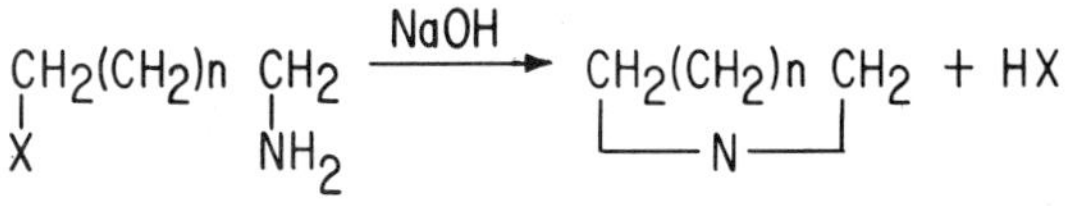

$$CH_2(CH_2)n\ CH_2 \xrightarrow{\ NaOH\ } CH_2(CH_2)n\ CH_2 + HX$$

Salomon and Freundlich (25, 59, 60) have extensively studied the velocities of formation of cyclic imines. While the results were considered inaccurate by Stoll (66, 67, 68), they at least provide a basis for comparisons. Clearly, greater energy of activation is required to form three- and four-membered rings than five- and six-membered imines. In ring-fission reactions involving cyclic imines it was found that the energy of activation of unstrained pyrrolidine was greater than for the highly strained ethylene imine.

The studies of Freundlich and Salomon (25) clearly show, however, that these reactions can be strongly influenced by extraneous factors such as the presence of substituent groups. Ring closure proceeds more than 1000 times faster for β-phenyl-β-chloroethylamine than for unsubstituted β-chloroethylamine.

Structural Problems

Baeyer's theory has frequently been valuable in settling structural questions since it is possible by applying the theory to rule out certain structures or conversely to accept them in lieu of more strained alternatives.

Blanc's rule (7, 8), which was introduced in 1907, holds that when succinic and glutaric acids are heated with acetic anhydride and then distilled at about 300° C., they yield the corresponding anhydride whereas adipic and pimelic acids under similar conditions yield cyclic ketones. In each case, the product formed is a five- or six-membered ring where strain is minimal. The rule took on practical importance, not only in distinguishing 1,4- and 1,5-dicarboxylic acids from their 1,6 and 1,7 counterparts but in determining ring sizes. Following the introduction of a double bond into the ring, it may be opened and oxidized to a dicarboxylic acid, heated, and distilled. A six-membered ring yields a cyclic ketone; a five-membered ring yields the anhydride.

The rule can be misleading, however, as in the case of cholesterol. When its structure was being worked out it was believed for some time, largely on the basis of the Blanc rule, that the B ring was five-membered.

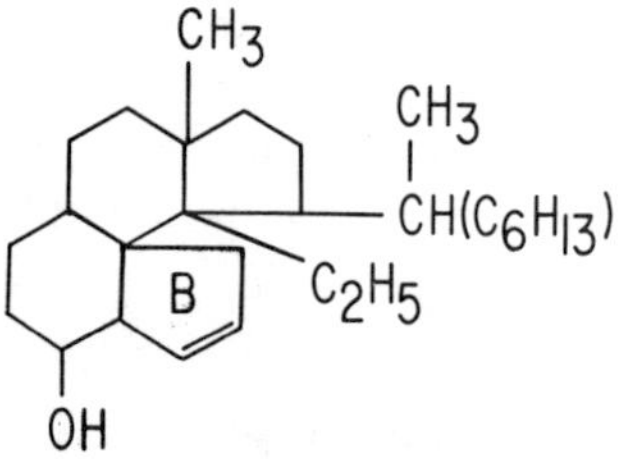

ORIGINAL CHOLESTEROL STRUCTURE 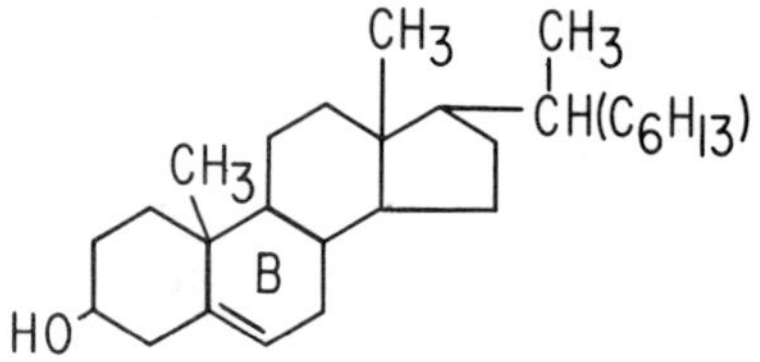   REVISED CHOLESTEROL STRUCTURE

It was ultimately shown by Windaus that substituted adipic acids can give the anhydride rather than a cyclic ketone (73). It had become known that substituents on a chain frequently facilitated ring closure. For example, β,β-dimethylglutaric acid forms an anhydride more easily than glutaric acid; α, α-dimethyladipic acid forms the ketone more readily than adipic acid. Thus the suspicion might at least be enter-

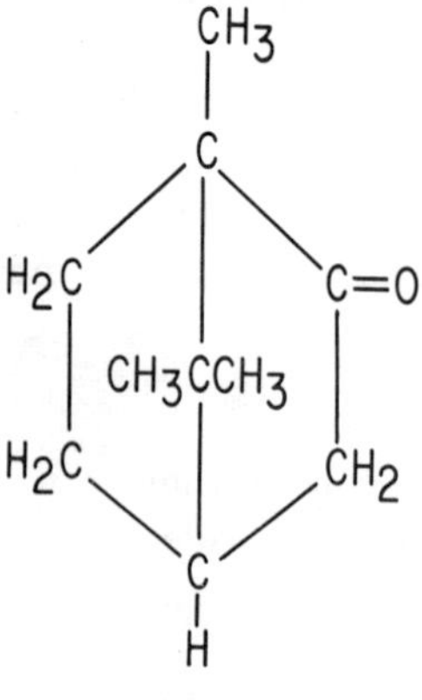

CAMPHOR 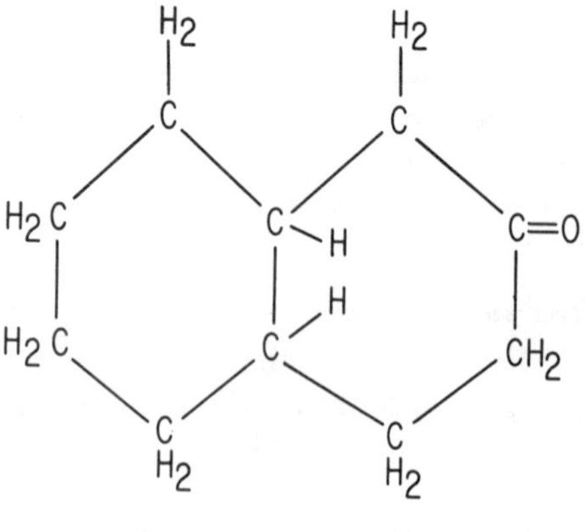   CIS–β – DECALONE

tained whether a substituted adipic acid might even form the anhydride. This is exactly what was happening in the B ring of cholesterol when Blanc's rule was applied.

Strain theory proved particularly valuable in the field of terpene chemistry where bridged rings are frequently encountered. For example, camphor has two nonequivalent asymmetric carbon atoms; hence, theory would predict four stereoisomers. Actually only one pair of enantiomorphs is known—the form having the cis configuration. The trans forms would be highly strained and hence are unknown. Construction of a model of *cis*-camphor reveals that even here, strain must be significant. This appears to be borne out by heat of combustion which is 1410.7 kcal. per mole (51), compared with 1400.1 in the comparatively strain-free molecular isomer, *cis*-β-decalone (31).

Bredt's rule (11) suggests that ring atoms connected by a bridge cannot also participate in a double bond. This rule excludes the existence of compounds like the one shown below since such molecules would

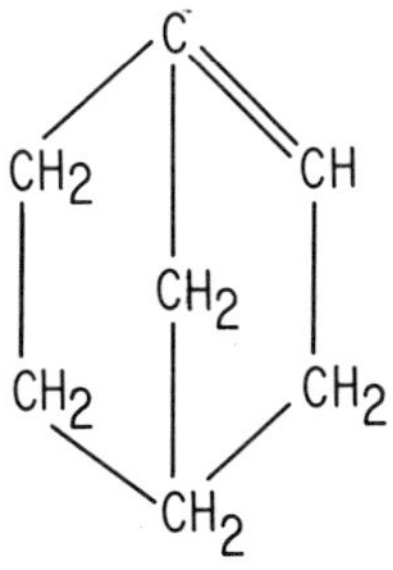

result in severe strains. However, Prelog and his associates (48, 49) have demonstrated that the rule loses significance provided the rings become sufficiently large as in the following compound where n equals eight or more.

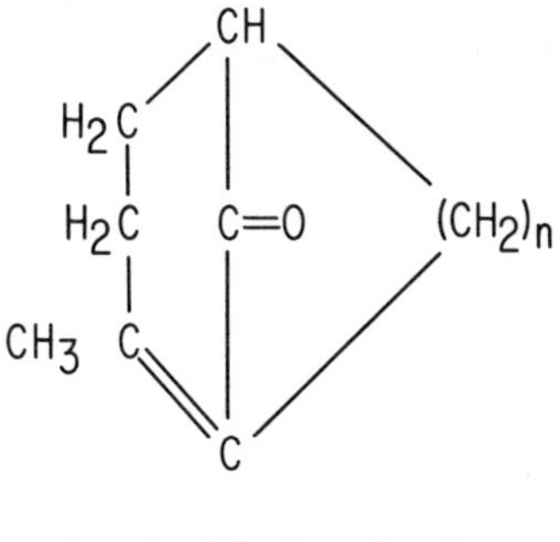

The mere insertion of a double bond into a ring can create problems involving strain. Cyclohexene, for example, exists only in the cis

form. With increasing ring size the strain decreases, and *trans*-cyclo-octene was prepared successfully in 1953 in Cope's laboratory (*14, 15*). Heats of hydrogenation in acetic acid solution revealed values of 32.24 kcal. per mole for the trans compound and 22.98 kcal. per mole for the cis form (*14, 15*). With larger rings the strains decrease, but that of the trans form remains comparatively greater.

Rings containing triple bonds would be expected to show excessive strains until a ring of fairly large size is present. The smallest known cycloyne is cyclooctyne, prepared in 1938 by Domnin (*9, 20*). Sondheimer and his associates (*62, 63*) have prepared ring compounds containing four and more triple bonds in which strain does not appear to be significant. Their simplest molecule is cyclohexa-1,3,9,11-tetrayne.

In the case of aromatic compounds the possibility of joining the opposite carbon atoms in the benzene ring by a bridge is greatly decreased since the benzene ring is planar rather than puckered. Hence, the para positions can be joined only when the bridge is fairly long. A successful syntheses was made by Spanagel and Carothers (*64*) in 1935.

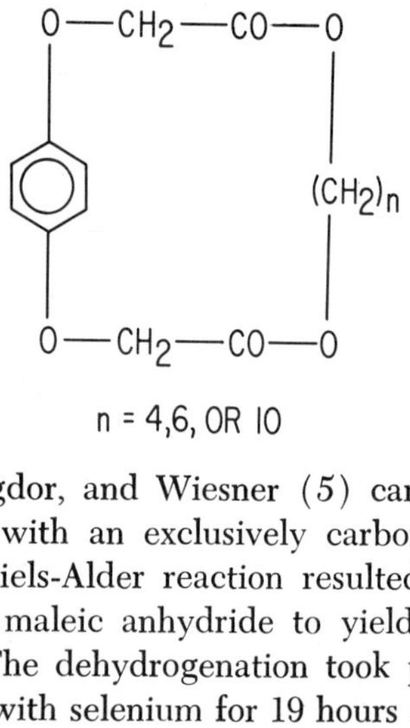

$$n = 4, 6, \text{OR } 10$$

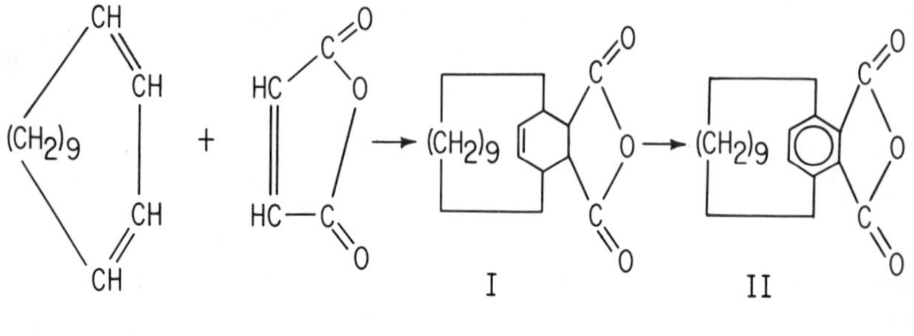

In 1952 Bartlett, Figdor, and Wiesner (*5*) carried out a synthesis leading to a compound with an exclusively carbon bridge connecting the para positions. A Diels-Alder reaction resulted in the addition of cyclotrideca-1,3-diene to maleic anhydride to yield I, which was then dehydrogenated to II. The dehydrogenation took place with difficulty, requiring I to be heated with selenium for 19 hours at 370°C. When the analog of I containing 12 instead of 13 methylene groups in the bridge was prepared, it proved impossible to dehydrogenate to the aromatic form. Analogs containing 14 or more methylene groups were dehydrogenated in a few minutes over palladium at 340°C, revealing that strain was apparently no longer of any consequence.

Strain has also been studied by Cram and associates (*17*) in paracyclophanes, where two benzene rings are linked together by two saturated chains joined at the para positions as in III.

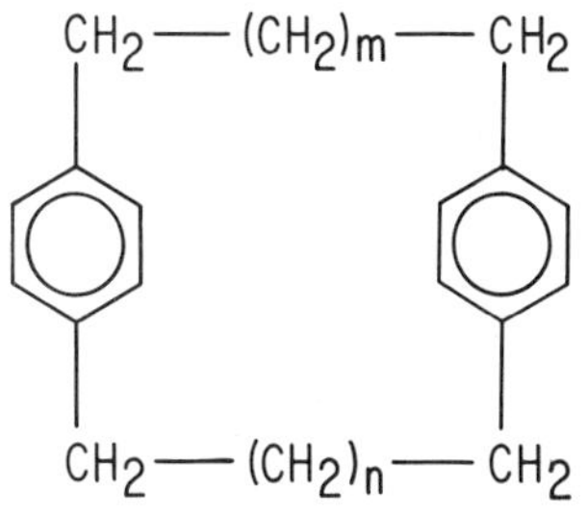

$$CH_2 \longrightarrow (CH_2)_m \longrightarrow CH_2$$

$$CH_2 \longrightarrow (CH_2)_n \longrightarrow CH_2$$

III

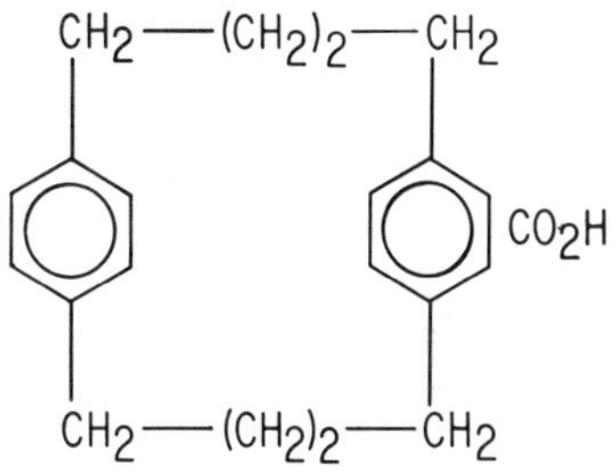

$$CH_2 \longrightarrow (CH_2)_2 \longrightarrow CH_2$$

$$CO_2H$$

$$CH_2 \longrightarrow (CH_2)_2 \longrightarrow CH_2$$

IV

The simplest compound they were able to prepare was the one in which $m = n = 2$, suggesting that strain is too great to permit simpler analogs to exist. Brown (*12*) showed that even in this compound the benzene rings are not planar, but boat shaped, revealing considerable strain. Further, Cram's group showed that compound IV can be resolved, revealing that rotation about the single bonds is restricted. Analogs where m and n are greater than 2 show greater freedom of the aromatic rings from one another as revealed by reactivity and by ultraviolet absorption spectra. For compound III when $m = n = 2$, for example, the ultraviolet spectrum (per aromatic ring) is clearly different from that of an unstrained paradialkylbenzene. When m and n become larger, this difference disappears.

Thermochemical studies of multiple ring systems have been useful in elucidating bond strains. Attention has already been called to the higher heat of combustion of camphor over its molecular isomer, *cis-β*-decalone. Comparisons of this sort are open to the criticism that the molecules are otherwise dissimilar because one is compact while the other is open. However, the comparisons are consistent with the strain observed in molecular models and with certain other criteria and hence are worth considering.

Following Mohr's prediction (*39*) of two isomers of decalin, Hückel (*27, 28*) succeeded in preparing them. Such isomers had been considered too strained to exist by those chemists who believed in planar rings.

Hückel and Friedrich (*31*) went on to determine the heats of combustion which were 1499.9 kcal. per mole for *cis*-decalin, 1497.1 for the trans form. They obtained, for the isomeric decalones, values of 1402.3 and 1400.1 kcal. per mole for *cis*- and *trans*-β-decalone respectively. These differences are not great but are consistent with the strains predictable from models.

Barrett and Linstead (*4*) prepared the two bicyclooctanes and found the heat of combustion of the trans form to be about 6 kcal. per mole greater than that of the cis form. The corresponding β-ketones gave a similar result.

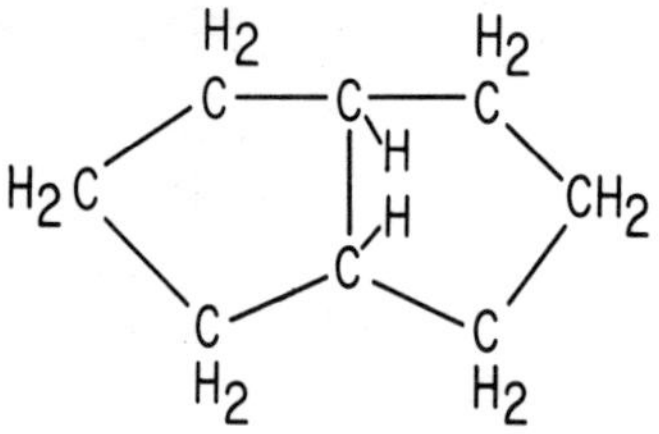

CIS—BICYCLOOCTANE 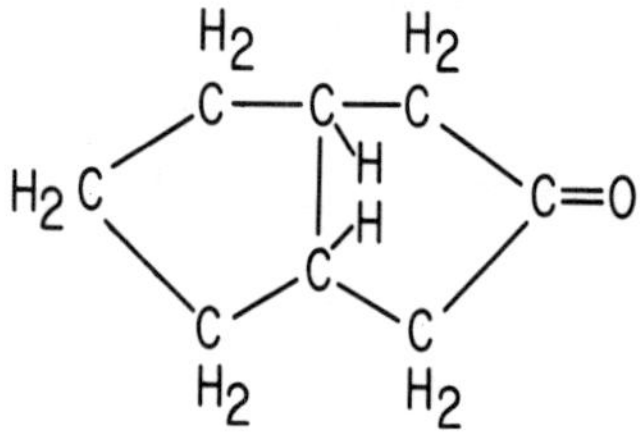   CIS−β−BICYCLOOCTANONE

Convincing evidence of strain was observed in bridge-ring systems by Alder and Stein (*1, 2*) in their comparisons of heats of combustion of 2,5-*endo*-methylenecyclopentanone (I) and 2,5-*endo*-ethylenecyclo-

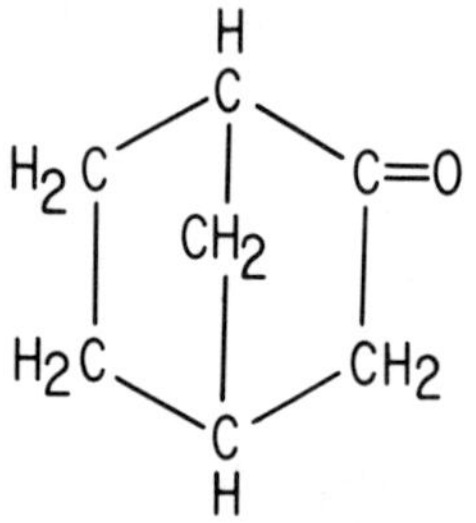

I   II
HEAT OF COMBUSTION   HEAT OF COMBUSTION
OF 946.1 KCAL./MOLE   OF 1096.8 KCAL./MOLE

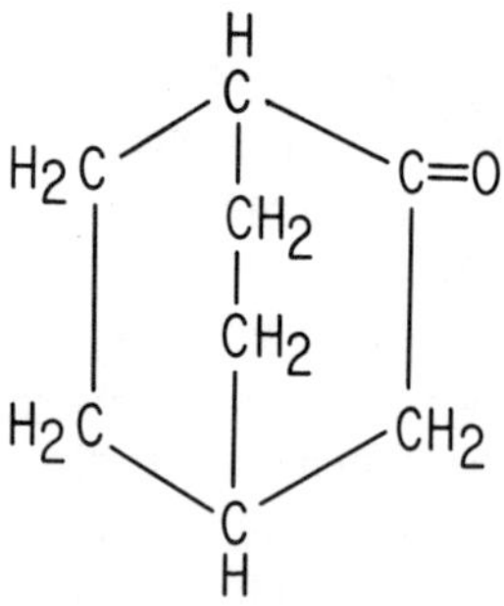

pentanone. The difference of 150.7 kcal. falls significantly short of the expected 157 kcal. for an added methylene group. However, construction of models predicts a large amount of strain in I whereas II is essentially strainless.

Quantum Mechanics

Quantum mechanics has been useful in shedding light on certain questions raised by Baeyer's strain theory. First, the Baeyer paper treated the double bond as a highly strained two-membered ring where the strain amounted to $+54°$ $44'$. The high degree of reactivity at double bonds was considered consistent with this point of view. However, the double bond has come to be treated as a combination of a sigma bond and a pi bond. The presence of the pi bond satisfactorily accounts for the reactivity without taking exceedingly high strain into account; hence strain theory drops out of consideration in connection with theory of double bonds. The triple bond is similarly excluded from strain consideration since quantum mechanics treats it as a combination of one σ bond and two π bonds. It is ironic that the main part of Baeyer's paper (*3*) dealt with compounds containing acetylenic linkages.

In the case of cyclopropane and cyclobutane, there is genuine strain in the Baeyer sense. Coulson and Moffitt (*16*) have sought to deal with this through the concept of "bent" bonds. Studies on bond angles have revealed that the normal tetrahedral angle of $109°$ $28'$ is to be expected only when four identical groups are attached to a carbon atom. When the groups differ, the bonds are no longer equivalent and point to the corners of an irregular tetrahedron. Since the bond angle between valencies cannot be less than $90°$ (in the case of pure p-orbitals), the calculated angle of $60°$ in cyclopropane is untenable. Coulson argues that the smallest carbon bond angle to be expected is $104°$; hence, in cyclopropane the hybridized orbitals cannot point directly toward one another, and loss of overlap results in "strain" or weakness of the bonds (Figure 7). In the case of cyclobutane a similar but less extensive loss of overlap occurs, and weakness of the ring is somewhat less apparent.

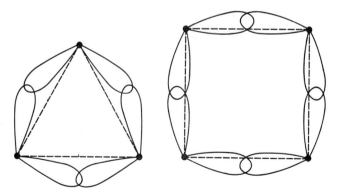

Figure 7.   "Bent" bonds of cyclopropane and cyclobutane.
Drawn after description by Coulson and Moffitt (16)

In cyclopentane there is no need for bent bonds, and in higher cyclo-paraffins the puckered conformation eliminates strain. It is to be sus-pected that "bent" bonds of the sort described would be found in compounds such as the bridge-rings where strain is frequently con-siderable.

Summary

During the 80 years since its introduction, the Baeyer strain theory has been an important conceptual tool in the development of chemistry. Actually, it represents an extension of the concept of the tetrahedral carbon atom introduced in 1874 by van't Hoff. Frequently, as in the case of the terpenes and the steroids, the strain concept has been a sig-nificant aid in resolving structural problems. At the same time, it has sometimes misled but generally only when chemists read into the theory certain rigidities which were misleading, as for example the idea that rings must be planar. Once the concept of the puckered ring was under-stood, many structural ambiguities could be resolved.

Physical evidence, such as that available from heats of combustion, dipole moments, and spectral data, has tended to support the evidence of strain suggested by molecular models. Chemical evidence connected with equilibrium constants, reaction velocities, and stability of compounds has likewise supported the theory. Such evidence has not always been as abundant as one might wish, but that which is available generally tends to support Baeyer's concept.

Even the notion of nonexistence of three- and four-membered rings in nature has had to be abandoned in the face of the discovery of four-membered rings in truxillic and truxinic acids in coca leaves and of three-membered rings in the pyrethrins and in sterculic and lactobacillic acids.

As is generally true of a theory in long use, certain aspects of it have had to be abandoned, and supplementary concepts have become necessary. Thus, the intense strain associated with double and triple bonds has been largely abandoned in favor of σ and π bonds. On the other hand, strain theory originally took no account of the influence of neighboring atoms and groups in ameliorating or intensifying strain. However, such supplementary concepts have not forced revision of the fundamental assumptions laid down by Baeyer.

Acknowledgment

I am indebted to K. T. Finley for calling to my attention the work of Deutsch and Buchman (19).

Literature Cited

(1) Alder, K., Stein, G., *Ber.* **67**, 613 (1934).
(2) Becker, G., Roth, W. A., *Ibid.*, **67**, 627 (1934).
(3) Baeyer, A. von, *Ber.* **18**, 2277 (1885). For an English translation of all but the last paragraph of this paper *see* Leicester, H. M., Klickstein, H. S., "A Source Book of Chemistry," p. 465, McGraw-Hill Book Co., New York, 1952.
(4) Barrett, J. W., Linstead, R. P., *J. Chem. Soc.* **1935**, 436, 612.
(5) Bartlett, M. F., Figdor, S. K., Wiesner, K., *Can. J. Chem.* **B30**, 291 (1952).
(6) Bernthsen, A. "Kurtzes Lehrbuch der organischen Chemie," p. 359, F. Vieweg, Leipzig, 1918.
(7) Blanc, H. G., *Compt. Rend.* **144**, 1356 (1907).
(8) Blanc, H. G., *Bull. Soc. Chim. France* **3**, 778 (1908).
(9) Bloomquist, A. T., Liu, L. H., *J. Am. Chem. Soc.* **75**, 2153 (1953).
(10) Böeseken, J., Griffen, J. von, *Rec. Trav. Chim.* **39**, 183 (1920).
(11) Bredt, J., *Ann.* **395**, 26 (1913).
(12) Brown, C. J., *J. Chem. Soc.* **1953**, 3265.
(13) Buchman, E. R., Reims, A. O., Schlatter, M. J., *J. Am. Chem. Soc.* **64**, 2703 (1942).
(14) Cope, A. C., Pike, R. A., Spencer, C. F., *J. Am. Chem. Soc.* **75**, 3212 (1953).
(15) Cope, A. C., Moore, P. T., Moore, W. R., *Ibid.* **81**, 3153 (1959).
(16) Coulson, C. A., Moffitt, W. E., *Phil. Mag.* **40**, 1 (1949).
(17) Cram, D. J., Allinger, N. L., *J. Am. Chem. Soc.* **77**, 6289 (1955).
(18) Derx, H. G., *Rec. Trav. Chim.* **41**, 312 (1922).
(19) Deutsch, D. H., Buchman, E. R., *Experientia* **6**, 462 (1950).
(20) Domnin, N. A., *J. Gen. Chem. U.S.S.R.* **8**, 851 (1938); *C. A.* **33**, 1282 (1939).
(21) Finley, K. T., *J. Chem. Ed.* **42**, 536 (1965).
(22) Fittig, R., Roeder, F., *Ber.* **16**, 372 (1883).
(23) Fittig, R., Roeder, F., *Ann.* **227**, 13 (1885).
(24) Freund, A., *Monatsh.* **3**, 625 (1882).
(25) Freundlich, H., Salomon, G., *Ber.* **66**, 355 (1933).
(26) Goss, F. R., Ingold, C. K., *J. Chem. Soc.* **127**, 2776 (1925).
(27) Hückel, W., *Ann.* **441**, 1 (1925); **451**, 109 (1926).
(28) Hückel, *Ber.* **58**, 1449 (1925).
(29) Hückel, W., "Der gegenwartige Stand der Spannungs Theorie, Fortschritte der Chemie, Physik und Physikalische Chemie," Ser. A, Bd. 19, Heft 4, A. Eucken, ed., Gebrüder Borntraeger, Berlin, 1927.
(30) Hückel, W., "Theoretical Principles of Organic Chemistry," 2 vols., transl. by F. H. Rathmann, Elsevier Publ. Co., Amsterdam and New York, 1955, 1958.
(31) Hückel, W., Friedrich, H., *Ann.* **451**, 132 (1926).
(32) Hückel, W., Gercke, A., Gross, A., *Ber.* **66**, 563 (1933).
(33) Kaarsemaker, S., Coops, J., *Rec. Trav. Chim.* **71**, 261 (1952).
(34) Knowlton, J. W., Rossini, F. D., *J. Res. Natl. Bur. Std.* **43**, 113 (1949).
(35) Lipp, A., *Ber.* **18**, 3280 (1885).
(36) Markovnikov, W. and Krestovnikov, A., *Ann.* **208**, 334 (1881).
(37) Meyer, V., *Ann.* **180**, 192 (1875).
(38) Meyer, V., Jacobson, P., "Lehrbuch der Organischen Chemie," Vol. 2, no. 1, p. 5, Veit, Leipzig, 1902.
(39) Mohr, E., *J. Prakt. Chem.* **98**, 315 (1918).
(40) Perkin, W. H., Jr., *Ber.* **16**, 208 (1883).
(41) Perkin, W. H., Jr., *Ber.* **16**, 1793 (1883).

(42) Perkin, W. H., Jr., *Ber.* **17**, 54 (1884).
(43) Perkin, W. H., Jr., *Ber.* **18**, 3246 (1885).
(44) Perkin, W. H., Jr., *Ber.* **19**, 2557 (1886).
(45) Perkin, W. H., Jr., *J. Chem. Soc.* **65**, 86 (1894).
(46) Perkin, W. H., Jr., *J. Chem. Soc.*, **1929**, 1347.
(47) Perkin, W. H., Jr., Haworth, E., *J. Chem. Soc.* **73**, 330 (1898).
(48) Prelog, V., Wirth, M. M., Ruzicka, L., *Helv. Chim. Acta* **29**, 2425 (1946).
(49) Prelog, V., Barman, P., Zimmerman, M., *Ibid.* **32**, 1284 (1949).
(50) Richter, V. von, Anschütz, R. "Chemie der Kohlenstoffverbindungen," vol. 2, p. 3, Fr. Cohen, Bonn, 1913.
(51) Roth, W. A., Östling, G. J., *Ber.* **46**, 313 (1913).
(52) Ruzicka, L., *Chem. Ind.* **1935**, 2.
(53) Ruzicka, L., *Helv. Chim. Acta* **9**, 230, 715, 1008 (1926).
(54) Ruzicka, L., *Bull. Soc. Chim.* **43**, 1145 (1928).
(55) Ruzicka, H., Stoll, M., Schinz, H., *Helv. Chim. Acta.* **9**, 249 (1926).
(56) Ruzicka, L., Brugger, W., Pfeiffer, M., Schinz, H., and Stoll, M., *Ibid.* **9**, 499 (1926).
(57) Sachse, H., *Ber.* **23**, 1363 (1890).
(58) Sachse, H., *Z. Physik. Chem.* **10**, 203 (1892).
(59) Salomon, G., *Helv. Chim. Acta* **16**, 1361 (1933); **17**, 851 (1934); **19**, 743 (1936).
(60) Salomon, G., *Trans. Faraday Soc.* **32**, 153 (1936); **34**, 1311 (1938).
(61) Sebelius, H., "Zur Kentniss der Lactone, namentlich ihrer Hydrolyse," Inaugural Dissertation, Lund, 1927. *See* summary in Hückel, W., "Theoretical Principles of Organic Chemistry," Vol. 2, pp. 892-896, Elsevier, Amsterdam, 1958.
(62) Sondheimer, F., Amiel, Y., *J. Am. Chem. Soc.* **78**, 4178 (1956).
(63) Sondheimer, F., Amiel, Y., Wolovsky, R., *Ibid.* **79**, 4247 (1957).
(64) Spanagel, E. W. Carothers, W. H., *J. Am. Chem. Soc.* **57**, 935 (1935).
(65) Staudinger, H., Muntwyler, O., Ruzicka, L., Seibt, S., *Helv. Chim. Acta* **7**, 401 (1924).
(66) Stoll, M., *Helv. Chim. Acta* **19**, 748 (1936).
(67) Stoll, M., *Trans. Faraday Soc.* **32**, 153 (1936).
(68) Stoll, M., Rouvé, A., *Helv. Chim. Acta* **19**, 1079 (1936).
(69) Turner, R. B., Meador, W. R., *J. Am. Chem. Soc.* **79**, 4133 (1957).
(70) Wasserman, A., *Helv. Chim. Acta.* **13**, 207 (1930).
(71) Werner, R., "Lehrbuch der Stereochemie," p. 353, Jena, 1904.
(72) Willstätter, R., *Ber.* **40**, 4459 (1907).
(73) Windaus, A., *Z. physiol. Chem.* **213**, 157 (1932).
(74) Wreden, F., Znatowicz, B., *Ann.* **187**, 163 (1877).

RECEIVED January 3, 1966.

Alternatives to the Kekulé Formula for Benzene: The Ladenburg Formula

VIRGINIA M. SCHELAR

University of Wisconsin, Madison, Wis.

The disposition of the fourth valence of the carbon atoms in benzene has caused extensive discussion and speculation. Following Kekulé's formula of 1865, a variety of formulas for benzene was proposed. Two trends were evident: the desire to arrive at the actual structure and the desire to devise formulas which were faithful to the functional behavior and broadly indicative of the structural relationship of the constituent elements. In 1869, Ladenburg criticized Kekulé's formula and suggested alternatives, one of which was the prism formula, which for a time was a serious rival of the hexagon. The strengths and weaknesses of the Ladenburg formula relative to the Kekulé formula are assessed. Recent laboratory studies by Viehe and co-workers have renewed interest in the prism structure.

O ne of the principal problems facing organic chemists after the development of structural theory in 1860 was the constitution of benzene and its derivatives. Some explanation for the numerous cases of isomerism was required. Since earlier analytical methods could not adequately establish the nature of a compound, chemists began to realize that synthesis was necessary in investigating the constitution of organic compounds. The first representation of an aromatic compound by a structural formula was given by Couper (*18*) (Figure 1). He represented salicylic acid as two groups of three carbon atoms to which other groups and atoms are attached, but he did not indicate the mode of linkage of the carbon atoms in each group. This formula was unsatisfactory.

---

[1] Present address: St. Petersburg Junior College, St. Petersburg, Fla.

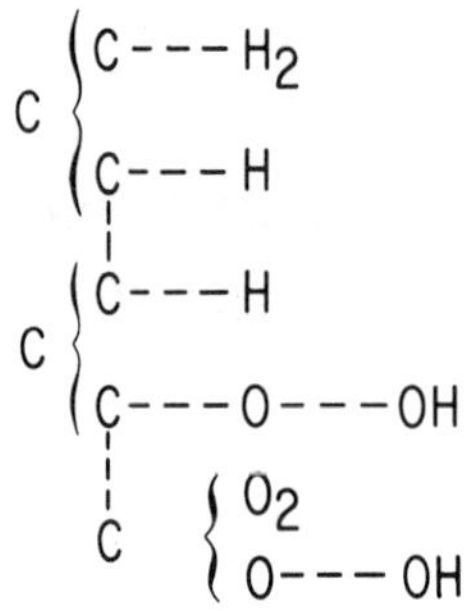

$$\text{C} \begin{cases} \text{C---H}_2 \\ \text{C---H} \end{cases}$$

$$\text{C} \begin{cases} \text{C---H} \\ \text{C---O---OH} \end{cases}$$

$$\text{C} \begin{cases} \text{O}_2 \\ \text{O---OH} \end{cases}$$

Figure 1. Structural formula of
salicyclic acid, after Couper (18)

Kekulé's Benzene Formulas

The properties of benzene are typical of aromatic compounds. From the beginning, the extreme stability of benzene and its marked tendency to react by substitution rather than addition were noted. Clearly these features were related to its physical structure. Kekulé published his first benzene formula in 1865 (28), using "roll symbols," by which the closed chain of carbon atoms was pictured as shown in Figure 2. The O repre-

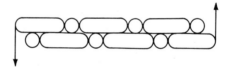

Figure 2. Kekulé's
"roll" formula

sents hydrogen, the ⊂⊃ represents carbon, and the arrows denote single carbon valences which saturate one another and make the configuration cyclic. The formula was not intended to indicate the actual structural arrangement of the carbon atoms in the ring.

In a second paper, which completes the essential part of his original theory (29), Kekulé proposed his well known hexagon formula for benzene (Figure 3). It was based on three assumptions: (1) the six carbon atoms in benzene form a closed chain or nucleus, (2) the molecule is symmetrical, and (3) each of the hydrogen atoms is united to one carbon atom. From this he argued for the existence of only one monosubstituted derivative and three diderivatives—conclusions for which rigorous proof did not exist. He preferred this formula to an alternative triangular distribution of the hydrogen atoms, which he also presented, but which he showed led to possibilities of isomerism in derivatives which were not

justified by facts. Neither formula accounted for the fourth unit of
valence of the carbon atoms in the nucleus. A short time later these
papers were republished in German, and the entire theory was again
published in the second volume of Kekulé's "Lehrbuch der Organischen
Chemie" (*30*).

*Figure 3.   Kekulé's
formula, 1865*

In 1866 Kekulé first published an explicit graphical representation
of the six carbon atoms in the nucleus (*31*) (Figure 4). He mutually
saturated the fourth carbon valences in pairs and gave a graphic formula
indicating alternate single and double bonds, which is equivalent to our
modern representation. The disposition of the fourth valence has caused
extended discussion and speculation. Kekulé said that no formula which

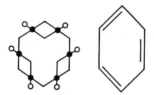

*Figure 4.   Kekulé's formulas,
1866. Black spheres represent carbon; white spheres
represent hydrogen*

arranges the atoms in one plane could completely express the linkings
of the atoms of carbon in the molecule of benzene (*33*). He said (*34*)
that the shortcomings would be removed if the:

. . . four units of affinity of the carbon atom, instead of being placed in
one plane, radiate from the spheres representing the atoms in the direction of hexahedral axes, so that they end in the faces of a tetrahedron—
A model of this description permits the union of 1, 2, and 3 units of
affinity, and, it seems to me, does all that a model can do.

Because of its greater simplicity and its direct applicability to most
of the problems concerning the relations of benzene derivatives, the
hexagon formula was used more than the tridimensional formula, of
which Figure 4 is a partial representation. Walker notes that after the

publication of his benzene theory, Kekulé wrote about 30 papers dealing with aromatic substances; in only about four of these did he use the benzene hexagon, and only to a limited extent (95).

Other Formulas

Since 1865 a variety of benzene formulas has been proposed, foremost of which are the following: the unsymmetrical formula of James Dewar (20) (Figure 5), the diagonal formula (Figure 6) of A. Claus (16), the prism formula of A. Ladenburg (41, 42, 51) (Figure 7), and the centric formula of H. E. Armstrong and A. Baeyer (1, 4) (Figure 8).

Figure 5.   Dewar's formula

Figure 6.   Claus' preferred formula and his alternative representation

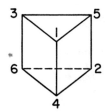

Figure 7.   Ladenburg's prism formula and his two alternative representations

Figure 8. Centric formula of Armstrong and Baeyer

Claus maintained (*14*), and not without cause, that Baeyer's view was identical with his own.

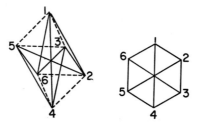

*Figure 9.   Thomsen's formula
with its plane projection*

*Figure 10.
Vaubel's formula*

*Figure 11.
Sachse's model*

○ is H   △ is C

*Figure 12.
Rosentiehl's formula*

In addition to the foregoing plane formulas, a number of stereochemical configurations are noteworthy: the octahedral formula of H. P. J. Julius Thomsen (*82*) (Figure 9), Vaubel's configuration (*90*) (Figure 10), the model of Sachse (*63*) (Figure 11), Rosenstiehl's representation

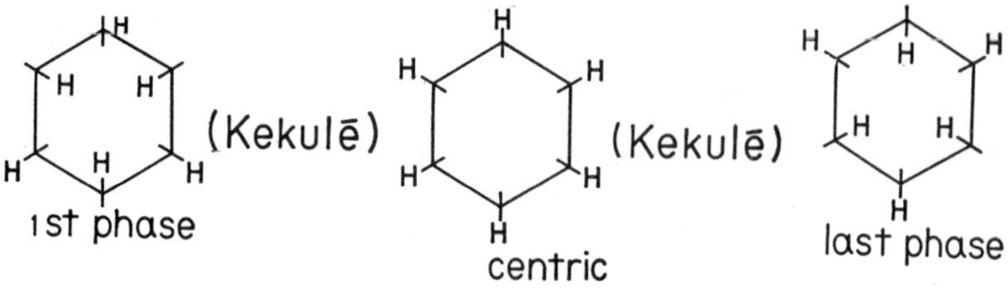

Figure 13.   *Plane projection of Collie's formulas. Kekulé's oscillation formulas would appear as intermediates on either side of the centric formula.*

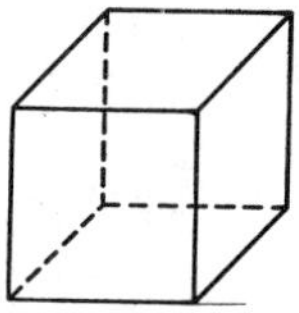

Figure 14.
König's formula

by six tetrahedra (62) (Figure 12), the dynamic formulas of J. N. Collie (15) (Figure 13), and the device of B. König (38) (Figure 14).

It is not possible to discuss here the characteristic features of the foregoing benzene formulas and their related theories. A careful study of their respective merits and demerits will show why none of the proposed formulas was accepted as complete. This is only a sample of the many proposed benzene formulas. Koerner, for example, in 1869 used a designation somewhat resembling that of Claus and Ladenburg (39) (Figure 15). However, he presumed that each atom of carbon was

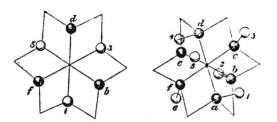

Figure 15.   Koerner's symbols

directly connected to three other atoms of carbon and the twelve atoms were arranged in four parallel planes. The atoms of hydrogen 1, 3, 5 and 2, 4, 6 are situated in the two extreme planes, and the carbon atoms a, c, e and b, d, f occupy the two intermediate planes. This symbol represents the six atoms of hydrogen as being of equal value and requires the existence of only three isomeric diderivatives. Moreover, the altogether different behavior of the 1:3 derivative as compared with the 1:2 and 1:4 derivatives is not only readily understood with the aid cf this symbol but might even be predicted.

Ladenburg

Efforts to determine the structure of benzene caused a great deal of "paper warfare." Two trends became evident: the desire to arrive at the actual structure and the desire to devise formulas which were faithful to the functional behavior and broadly indicative of the structural relationship of the constituent elements. The latter trend grew as it was realized that formulas were primarily expressions of behavior— i.e., based on observations of behavior. Ladenburg (Figure 16) clearly subscribed to the latter viewpoint as evidenced in his theory of aromatic compounds (52):

. . . the main purpose of structural formulas is found in a clear explanation of isomers, and this explanation frequently consists in showing that hydrogen atoms . . . are replaced by particular atoms or atom groups.

Figure 16.   Albert Ladenburg

His general approach is expressed as follows (51):
I am even of the opinion that in a science logical deductions arising from even a weak hypothesis, if they . . . can be confirmed directly by the facts, have their significance and remain even if their basis is shaken

or even shoved aside. Frequently a new form is then found for these, in which they become usable in other starting points.

Kekulé himself pointed out how the problem of the benzene constitution was to be solved (32):

It is only necessary to prepare, by methods as varied as can be devised, as great a number of substitution products of benzene as possible; to compare them very carefully with regard to isomerism; to count the observed modifications; and especially to endeavor to trace the cause of their differences to their modes of formation.

While it might seem from his memoirs of 1858 and 1866 that Kekulé regarded structural formulas as expressions of the real arrangements of the atoms in molecules, it should be recalled that he wrote (58):

"Rational formulae are decomposition formulae, and in the present state of science can be nothing more. These formulae give us pictures of the chemical nature of substance because the manner of writing them indicates the atomic groups which remain unattacked in certain reactions (the radicals), or lays stress on the constituents which play the same part in definite, oft-recurring metamorphoses (types). Every formula which expresses definite metamorphoses of a compound is rational; that one of the different rational formulae is the most rational which expresses the greatest number of metamorphoses."

It is instructive to compare Couper's view of rational formulae with that of Kekulé. Couper said:

"Gerhardt . . . is led to think it necessary to restrict chemical science to the arrangement of bodies according to their decompositions, and to deny the possibility of our comprehending their molecular constitution. Can such a view tend to the advancement of science? Would it not be only rational, in accepting this veto, to renounce chemical research altogether?"

Kekulé held that if the arrangement of atoms in molecules is ever to be understood, it will be by the study of physical properties rather than of chemical reactions. He thought it possible that one might thus attain to true constitutional formulae. Chemists do not now very strenuously dispute about the exact extent to which rational formulae express the molecular constitution of bodies. These formulae have proved themselves to be such powerful instruments of research that chemists are content to use them for the purpose in hand, without discussing what other purposes they may some day serve. The use of rational formulae is a representative instance of the fruitful employment of hypotheses for the advancement of accurate knowledge.

Various points in Kekulé's theory which at first were either fundamental assumptions or deductions from these have since been proved experimentally. Evidence for the symmetry of the molecule, from which it follows that there can be only one monoderivative, was supplied by the combined research of Hübner and Petermann, Wroblewsky and Ladenburg. Ladenburg proved the equivalence of the six hydrogen atoms in benzene (45). Hübner and Petermann and Wroblewsky proved that there is, relative to every hydrogen atom, a symmetrically situated ortho pair and a symmetrically situated meta pair (25, 97). The orientation of the substituents in the derivatives of benzene, merely indicated by Kekulé, has been successively treated by Baeyer, Graebe, Ladenburg, Greiss, and above all Koerner and Nolting, who replaced each of the six hydrogen atoms in turn by the —NH₂ group and found that the same compound—aniline—was obtained in each case (40, 59).

The Prism Formula

In 1869, Ladenburg who worked in Kekulé's laboratory criticized Kekulé's formula and suggested alternatives, among which was the prism formula, originally proposed by Claus. Although this formula could account for the various isomers of benzene, it excluded the possibility of the double bonds which must be present in dihydro- and tetrahydrobenzene. For a time the prism formula was a serious rival of the hexagon designation. For that reason and because of Ladenburg's many theoretical and practical contributions to the benzene theory, attention will be focused on the discussion centering on the relative merits of the Ladenburg and Kekulé formulas in the remainder of the paper.

Ladenburg, in his 1869 paper, used the proof of benzene's symmetry to attack a vulnerable point in Kekulé's formula, which represented the carbon atoms as linked by alternate single and double bonds. On this assumption, it follows that in the diderivatives two different ortho compounds, 1:2 and 1:6, should exist, and if the groups are different, two meta compounds also were possible. In other words, four diderivatives are possible. At most three have been obtained. Therefore, no such differences are observed. At first Kekulé ignored this difficulty, but continued discussion led him in 1872 to seek to meet the problem by his well known oscillation formula, in which the double and single bonds continually exchange places (35). [Much of the material contained in the earlier portion of Kekulé's paper (35) is fully discussed in ref. (96).] This was virtually a return to the simple hexagon of his original paper in which the distribution of the unsaturated valences is in effect ignored.

Kekulé believed that the view that there should be a difference between the 1:2 and 1:6 diderivative originated from the model used rather than from the ideas which this model insufficiently represents— i.e., it represents the configuration during one set of oscillations only. The difference between the two ortho positions was, in his opinion, more apparent than real, and he presented a hypothesis as to the way in which the atoms move in the benzene molecule. The atoms in molecules must be considered as being in continuous movement; no explanation was given as to the nature of the intramolecular motion. The most probable assumption appeared to be that the separate atoms of the system, possessing an essentially rectilinear motion, strike each other, and being elastic bodies, then recoil. What was then termed atomicity or equivalency acquired therefore a more mechanical significance: the equivalency became the relative number of contacts which occur in unit time between atoms. Applying this view to Kekulé's benzene formula, each carbon atom strikes two other carbon atoms during unit time—once against one and twice against the other. In the same unit of time each carbon atom also comes in contact once with the hydrogen atom, which during the same period makes one complete vibration. The number of contacts made by carbon atom 1 during the first two units of time are then: 2, 6, H, 2, 6, 2, H, 6, from which it is evident that each carbon atom strikes each of the two carbon atoms upon which it impinges an equal number of times—in other words, it bears the same relation to both contiguous atoms. The ordinary benzene formula represents only the contacts made during one unit of time. Thus it was, said Kekulé, that the false view had arisen that diderivatives in which the radicals occupy the positions 1:2 and 1:6, respectively, must be different. The explanation attempted by Kekulé was heavily criticized, particularly by Ladenburg, who devoted much study to substitution products of benzene and to supporting the prism formula which he adopted. In 1872 Ladenburg draws the conclusion from previous research that each carbon atom in benzene corresponds to two pairs of equivalent hydrogen atoms (44). Since the ordinary benzene formula of Kekulé does not suffice for this condition, some positive evidence is presented on behalf of the formula denoted by Ladenburg as the "second benzene hypothesis of Kekulé." In this paper Ladenburg briefly discusses the consequences of Kekulé's mechanical exposition of the mutual connections of the atoms. He distinguishes two separate points: (1) the nature of the notion of equivalence, and (2) the acceptance of one of two positions of unstable equilibrium of an atom in a molecule. It should perhaps be pointed out that Kekulé and Ladenburg agreed in rejecting the notion of variable valence.

Ladenburg's prism formula, like Claus' diagonal formula and Armstrong's centric formula, all agree with Kekulé's formula in connecting together six carbon-hydrogen groups by single bonds to form a closed chain. The disposal of the remaining six linkages is the point on which they differ. In the Claus and Ladenburg formulas the fourth unit of valence is represented as directly united to three other carbon atoms. In Ladenburg's prism formula the six carbon atoms are placed at the corners of a regular prism, the edges of which denote the linkages. The numbered positions correspond to the arrangement of the carbons in the hexagon, as determined by Koerner's principle of orientation. The diagonal corners of the prism faces are ortho, those occupying the ends of the vertical edges are para, and those at the corners of the triangular faces are meta positions. Ladenburg pointed out that a para-diderivative of benzene may be defined as one which can only give rise to a single triderivative, an ortho-diderivative is one which can furnish two isomeric triderivatives, and a meta-diderivative is one which can furnish three isomeric triderivatives—the two displacing radicals in the diderivatives being in each case alike (47). This definition was based entirely on facts and is in accord with today's use of the terms para, meta, and ortho as denoting 1:4, 1:3, and 1:2 compounds, respectively.

Addition Compounds

Kekulé objected to Ladenburg's formula on the grounds that it did not satisfactorily account for the formation of addition compounds. To convert benzene into cyclohexane, it was necessary to have recourse to the doubtful expedient of breaking one para and two meta linkages (Figure 17). Ladenburg admitted that the explanation of addition products was

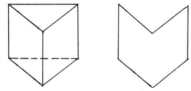

Figure 17.   Ladenburg conversion of benzene into cyclohexane

less elegant on the basis of his formula than Kekulé's, but he pointed out in 1869 that his formula agreed as well with the facts as did Kekulé's. Each gave a view of the formation of benzene from acetylene and the formation of mesitylene from acetone. Koerner, in the opening portion of his 1874 memoir, discusses the validity of the arguments upon which the then-accepted system of representing the constitution of benzene derivatives was founded. He notes that it had become customary to refer the diderivatives of benzene to the three phthalic acids, but the determination of the constitution of these bodies had its origin primarily in the determination of mesitylene and naphthalene. Inasmuch as the security of the entire system depended on determining these substances correctly, he questioned whether their constitution was well enough established to be used for all other aromatic derivatives and what degree of confidence could be placed in experiments in which acids containing eight atoms of carbon are connected with the more immediate derivatives of benzene containing only six. He believed that the only answer which could be given to these questions was decidedly negative. Thus with regard to mesitylene, although it appears extremely probable from its mode of formation that it is a symmetrically constituted trimethylbenzene, its production from three molecules of acetone at a comparatively high temperature, accompanied as it is by the elimination of three molecules of water, justified Koerner's doubt whether the reaction does not involve intramolecular change. He also objected to Graebe's conclusions concerning the constitution of phthalic acid. Koerner believed his experiments prove only that naphthalene may be regarded as built up of two benzene nuclei having two atoms of carbon in common. All speculation with regard to the relative positions of these two atoms was pure conjecture, and the conclusions as to the nature of phthalic acid depended entirely on the kind of symbol used to represent naphthalene—i.e., whether the two carbon atoms common to the two nuclei were assumed to be adjacent or nonadjacent.

One of the objections to the prism formula was that it did not express the well-known tendency of ortho compounds to form "inner anhydrides." Substitution phenomena in the benzene ring indicate that the ortho and para position have something in common which distinguishes them from the meta. Ladenburg's formula is not in accord with this since it implies a similarity between meta and para as distinguished from ortho. His formula represents the ortho-carbon atoms as not being directly connected, thus ignoring the analogy between the ortho position in benzene compounds and the alpha position in paraffinoid compounds and rendering the formulation of compounds like naphthalene and phenanthrene impossible—at least in accordance with the prevailing views. It may be well to recall exactly what Ladenburg did state (43) in his 1869 paper:

When I represent the constitution of benzene by an equilateral three edged prism, it signifies that I consider the 6 carbon atoms equal, that is the first hypothesis of Kekulé is sufficient. Since further the three-edged prism possesses two edges of different value, then each corner is not bound in exactly the same way with three others, with two namely

through edges, which at the same time bound a triangle and a square, with the third corner by a line, which only belongs to the side surface. It follows from this view that . . . 1:4 can not be like 1:2 and 1:6. The basis for this can be given . . .: two entering elements are bound to carbon at 1 and 2 or 1 and 6, which among one another stand in direct relation and moreover have a third carbon in common. The last condition is lacking in the position 1:4. The formula thus predicts 3 isomers in the substitution product C_6H_4AB, namely 1:2 = 1:6, 1:3 = 1:5, and 1:4.

That Ladenburg was concerned about this problem is obvious from two of his papers on condensations in the ortho group (48). He notes that the changes taking place in aromatic ortho groups, as observed in experiments then recently performed, are often different from those which occur in the other two groups, the removal of certain atoms sometimes leading to internal condensations for which no analogy was found in the meta and para groups.

Moreover, introducing two different groups into the prism formula gives rise to molecular asymmetry, which implies the existence of optical enantiomorphs. All attempts to resolve such compounds have been fruitless, and what is even more significant, there is no single instance of an optically active compound of benzene found in nature which owes its activity to the asymmetry of carbon in the nucleus rather than in a side chain. However, the stereochemical argument against the prism formula, although powerful, has little historical significance. Ladenburg did not attach any spatial significance to his formula. He states (42):

If one uses, as frequently happens, graphic formulas to illustrate constitution, geometrical relations for the mutual relations of the atoms are authoritative, at which we protest in the usual way, through the figure to intend to state the spatial positions.

van't Hoff's Formula

In 1881, van't Hoff criticized Ladenburg's formula on the grounds that the possibility of two ortho derivatives encountered in Kekulé's formula also exists in the prism formula (86). He numbered the prism formula (Figure 18) to correspond to the Kekulé formula and showed that in this the derivatives 1:2, 5:6 and 3:4 are perfectly alike but differ from 4:5, 2:3, and 6:1. It is impossible to turn the prism and match the first case to the latter. If two hydrogen atoms of benzene are replaced by two different groups (Figure 19), Kekulé's hexagon with fixed double bonds gives not only two ortho but also two meta derivatives. A product

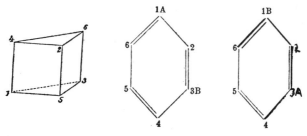

Figure 18.   van't Hoff's numbering

1:3 is different according to whether A or B enters the 1 position. However, Ladenburg's formula gives the same result. Consideration shows that I and II are absolutely different, a distinction which is not a consequence of the prism's position. van't Hoff concludes that since the same

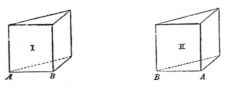

Figure 19.   van't Hoff's argument

difficulties beset the prism as the hexagon formulas—and on the whole Kekulé's hexagon is simpler and better adapted to explain facts than Ladenburg's prism—Kekulé's formula is preferable. Ladenburg pointed out that van't Hoff's error—i.e., the fact that there are two ortho and two meta derivatives possible using the prism formula—was caused by his considering only the position of the atoms in space and not with reference to their mode of connection (49). He reiterates his statement from his pamphlet (53):

Through the formula an account shall be given of structure, molecular weight, and the mode of union of the atoms.

This was in keeping with the opinion of the time. He stated that from this standpoint van't Hoff must acknowledge that the two formulas he gave are absolutely alike. He went on to point out that if van't Hoff's spatial viewpoint did not permit him to accept the identity of the two prism formulas, he should use the "David's cross" which Ladenburg presented along with the prism in 1869, and which for him was identical with the prism. The Ladenburg formula was essentially rejected before the full implications of the tetrahedral structure of carbon were generally accepted.

Emil Erlenmeyer gave a rather complete discussion of the benzene problem from the stereochemical standpoint in 1901 (21). In 1902, Carl Graebe also discussed various representations of the space formula of benzene and obtained a new figure (23), but Wilhelm Marchwald, in reference to Graebe's discussion, pointed out that all these configurations which do not contain the centers of gravities of their constituent atoms in the same plane may be *a priori* excluded on purely stereochemical grounds (55).

Support for the Prism Formula

The prism formula received apparent support in 1879 when Gruber, then Barth's assistant, found that protocatechuic acid when oxidized by nitric acid formed a new acid carboxytartronic acid, $HOC(COOH)_3$, which gave up carbon dioxide easily, yielding tartronic acid. Barth thought it improbable that the carboxy group had been cleaved in the process, and therefore the acid must have come from the benzene nucleus (6). This implied that in the benzene nucleus there was at least one carbon atom linked to three others. Consequently, from chemical evidence, they came to the same conclusion Julius Thomsen did by thermochemical research (72)—namely, that Kekulé's ring formula must be abandoned and replaced by the prism formula of Ladenburg. Kekulé later proved experimentally (36) that the so-called "carboxytartronic acid" was dibasic and not tribasic, that it is a dihydroxytartaric acid or tetrahydroxysuccinic acid since it could be prepared from tartaric acid. Therefore, this compound does not contain three carbon atoms directly united to each other. The formation of this substance readily follows from Kekulé's formula while considerable difficulties are met with when one attempts an explanation based on Ladenburg's representation.

"Trichlorophenomalic acid," was first discovered by Carius who represented it by an erroneous formula and various self-contradictory reactions. This formula was then apparently abolished by Krafft, but it was finally rehabilitated, explained, and used in support of the hexagon formula. Kekulé himself called this dramatic story a "comedy of errors." He showed that the formation of "trichlorophenomalonic acid" proved by him and O. Strecker (37) to be trichloroacetoacrylic acid, was more favorably explained by his formula than by Ladenburg's.

Nevertheless Ladenburg still believed that only the prism formula gave a clear idea of isomerism in the aromatic series. He considered that his approach was exactly the opposite of Kekulé's (53):

He [Kekulé] proceeded from a definite view of the manner of bonding of carbon atoms . . . in benzene, and derived from this certain conclusions about the equivalence of hydrogen atoms or the number of isomers. . . . [Ladenburg claims that in his pamphlet] what Kekulé placed at the start of his hypothesis, the manner of bonding of carbon in benzene, will be derived in an inductive way from the facts. . . .

Thermochemical Considerations

Ladenburg's formula furnished an accurate expression for the thermal relations according to Thomsen and for the molecular volume of benzene and its derivatives according to R. Schiff (64). According to Schröder (65) this also holds for the molecular refraction whereas according to Brühl (8) the opposite is the case. Brühl favored Kekulé's formula. Thomsen discussed the contradictory conclusions arrived at by Brühl from his investigations of molecular refraction as opposed to Thomsen's conclusions from his determination of the heat of combustion

(74). Thomsen pointed out how six carbon atoms mutually bound by three single bonds can affect the density in the same way and give the same molecular refraction as six carbons mutually bound by one single and one double bond. He states (75):

The supposed greater optical density, i.e. the about 6 units greater molecular refraction of benzene, can accordingly be as easily explained by the assumption of the presence of 9 single bonds, which bind each carbon to three others, as through the assumption of three double bonds. . . .

In fact, the early history of the use of physicochemical properties to decide whether double bonds are present in the benzene ring shows clearly that the predilictions of the observers apparently influenced the nature of the conclusion drawn from the same data. Smile's book (67) omits the heat of combustion from consideration. Because of this and because thermochemical considerations provided strong support for the Ladenburg formula in the period under consideration, the early discussions centering on this property will be considered in some detail.

In 1879 F. Stohmann criticized Thomsen's calorimeter and described a new one plus a correction factor for old values (69). This was a valid criticism and must be kept in mind when reviewing older values. H. P. J. Julius Thomsen in 1880 showed that the possible constitutional formulas of benzene could be arranged in nine groups (73).

Meyer

Lothar Meyer (56) obtained five possible formulas on the basis of the fact that the six hydrogens in benzene are equivalent and that with reference to one of the hydrogen atoms, the five others may occupy three different positions; therefore some of these occupy the same relative position to the first atom.

The only hypothesis capable of explaining these results is that the six atoms of hydrogen are equally distributed among the six carbon atoms and that the six pairs of atoms, CH, form a continuous circular chain. But as each carbon atom has three affinities available for union with the others, the chain need not of necessity form a single ring but may be a complicated, netlike structure. It is only necessary that all the carbon atoms should be linked together in exactly the same fashion. If the existence of free affinities is assumed, then this condition is only fulfilled by the hypothesis of a simple ring of six members; but in the case of double linking of the carbon atoms by the available affinities only two of the five possible ring-shaped arrangements of the atoms satisfy this condition. Let C^1, C^2, C^3, C^4, etc., indicate the six carbon atoms, and let the corresponding indices at the foot of each C denote the carbon atoms with which it is directly united, then the five following combinations are obtained. These formulae are not intended to indicate that the atoms are arranged in a circle in space, but only to show which atoms are in direct union with each other. For instance, the relative position of the

I.	II.	III.	IV.	V.
C^1_{622} C^2_{131}	C^1_{622} C^2_{131}	C^1_{622} C^2_{131}	C^1_{623} C^2_{135}	C^1_{624} C^2_{135}
C^3_{244} C^4_{353}	C^3_{245} C^4_{356}	C^3_{246} C^4_{355}	C^3_{241} C^4_{356}	C^3_{246} C^4_{351}
C^5_{466} C^6_{515}	C^5_{463} C^6_{514}	C^5_{461} C^6_{513}	C^5_{462} C^6_{511}	C^5_{462} C^6_{513}

or, represented graphically:

atoms in space may be the same as the angles of a regular octahedron; the affinities uniting the atoms would then act in the direction of nine of the twelve edges of the octahedron. Of the five possible combinations only the first and the two last satisfy the condition of all the carbon atoms' being linked together in precisely the same fashion.

Of Thomsen's nine groups, only two agree with the chemical characteristics of benzene—i.e. the three isomeric triderivatives. These two are Kekulé's hexagon formula, with alternate single and double bonds, and Ladenburg's formula, in which the six carbon atoms are joined through nine single bonds and each carbon atom is bound to three other carbon atoms.

Thomsen's Calculations

In 1880 (72) Thomsen proposed to decide between these formulas on the basis of his theory regarding heats of combustion and formation of hydrocarbons. By his previous research he had shown that for single and double bonds an equal amount of energy is developed while for a triple bond the amount of energy developed is nil. He explained that if the energy developed by the single bond is r, the energy developed by the second "affinity" is zero, and the energy developed by the third affinity is $-r$, then the double bond will give an energy of r, and a triple bond will give nil. This result was in accord with the chemical character of the carbon compounds. He had shown that the splitting-off of each "affinity" in paraffins is associated with a heat absorption of 14,570 heat units. In unsaturated compounds the multiple bonds form the attacking point for chemical reagents. For instance, when a molecule of chlorine acts on a molecule of ethylene, the double bond is converted into two single bonds, and the chlorine combines with both carbon atoms with all the energy which corresponds to the "affinity" between chlorine and carbon. The conversion of the double bond into two single bonds is not accompanied by an alteration of energy. In the case of acetylene, energy is developed in addition to that developed by the "affinity" of chlorine for carbon. However, when chlorine acts on a paraffin, there must be a splitting-off of a bond and an absorption of 14,570 units, whether the reaction consists of the expulsion of a hydrogen atom or a dissociation into two hydrocarbon radicals. On the basis of this hypothesis, double bonds can be converted into single bonds by very violent reactions while in all feeble reactions there will only be a substitution of the hydrogen atoms. The stability of benzene points to the absence of double bonds.

According to Thomsen's previous research (73), the heat developed in the formation of a hydrocarbon at constant volume can be expressed by the formula, $(C_nH_{2m}) = -nd + (2m + x + y)r$, in which d is the dissociation heat of carbon (39,200 heat units), x and y are the number of single and double bonds, and r the heat absorption in the combination of two carbons, or one carbon and one hydrogen = 14,570 heat units. For benzene, adopting Kekulé's hypothesis, $x = 3$, $y = 3$, $2m = 6$; for Ladenburg's hypothesis, $x = 9$, $y = 0$, $2m = 6$. The heat of formation of gaseous benzene will be —60,360 heat units in the first case and —16,650 heat units in the second. He used a value of 1,160 heat units greater for the heat of formation of benzene at constant pressure. The above values then become —59,200 and —15,490. If these values are subtracted from the heats of combustion of the constituents of benzene (786,840), the heat of combustion of benzene vapor becomes, 846,040 for the first hypothesis and 802,230 heat units for the second. The experimental value found was 805,800 heat units. The agreement between the experimental value and that theoretically deduced by adopting Ladenburg's hypothesis led Thomsen to conclude that "the six carbon atoms of benzene are combined together by nine single bonds, and the hypothesis adopted hitherto of a constitution of benzene with three single and three double bonds is not confirmed by experiment" (76, 77). (In a paper (76) on the heat of combustion of benzene, Thomsen considered the heat of formation to be equal to the difference between the heats of combustion of constituents of the compound and the compound itself and deduced the values used above as follows: Heat of combustion of benzene vapor (experimental) = 805,800 h.u. Heat of combustion of constituents: (CO_2) = 96,960 h.u.; (H_2O) = 68,360; total 786,840 h.u. Heat of formation of gaseous benzene at constant pressure = —18,960 h.u. Heat of formation of gaseous benzene at constant volume = —20,120 h.u.)

In papers on benzene and dipropargyl he reaffirms his support of a representation with nine single bonds (78, 79). In the 1882 paper of this series, he corrected his value for the heat of combustion of benzene to 787,950 cal. He attributed his former error to a mistake in calculations and impure benzene.

The discussion between Thomsen and Brühl continued. In 1881 Brühl sought to connect the changes of refractive power and the heat of combustion of organic compounds (9). Thomsen, in 1882, sought to prove that the apparent connection between the specific refractive power and the heat of combustion arose principally from the change in molecular weight on oxidation (*80*). He concluded (*81*) that "the qualitative change in the heat of combustion of the substance through oxidation or substitution forms a completely unusable basis for an investigation of the connection of optical and thermal properties."

Brühl published a criticism of attempts to relate thermochemical values and the chemical constitution of compounds after Thomsen proposed his own structural formula in 1886 (*10, 11*). In an 1894 paper, Brühl noted that no large changes in molecular volume and refraction constants are observed in the passage of benzene to dihydrobenzene as would be expected if the molecular structure of benzene materially differed from that of its dihydroderivative (*12*). He concluded that the hypothesis that benzene possesses a cyclic or a diagonal constitution was therefore opposed to the facts. Claus criticized Brühl's paper (*15*) and was in turn criticized by Marckwald (*54*), who also considered the deductions which Stohmann drew from his calorimetric determinations (*70*) concerning the constitution of benzene as premature. However, by this time interest was no longer centered on the Ladenburg formula as an alternative to the Kekulé formulation.

The discussion by Horstmann of the various results obtained in determining the heats of combustion of various saturated and unsaturated hydrocarbons, including benzene, deserves mention (*24*). He concludes that in physical properties benzene is intermediate between the saturated and unsaturated compounds—a conclusion which Baeyer had already drawn from a study of the chemical properties.

Re-evaluation of Thermochemical Data

To illustrate the inability of chemists to agree on the significance of thermochemical data, several papers written in 1912, which sought to reinterpret old data rather than contribute any new experimental results, are of some interest. Willebrord Tombrock reviewed the problem of benzene structure from the thermochemical standpoint (*84*). He called attention to the fact that the amounts of heat developed in subsequent hydrogen additions to benzene are not the same but differ considerably, and they are less than the heat of addition in open chain compounds. The diminution of the ordinary addition heat developed was ascribed to the absorption of energy in the molecule, part of the true addition energy then being retained. He assumed that this retained energy serves to cause a kind of tension or strain which is possible in a closed-chain structure. He develops an equation from which he thinks the true addition heat for benzene may be calculated. This equation was to clarify the structure of benzene since the calculated heat of combustion depends on the structure assumed for benzene. He considered first Baeyer's centric formula with nine single bonds and calculated 792.72 for the heat of combustion from Thomsen's values for bonds. Using Thomsen's experimental value of 799.35 and substituting in an equation derived by himself, he obtained the addition heat when the amount of energy necessary to open the benzene ring was determined independently—i.e. by subtracting the calculated heat of combustion of benzene from the experimental value. His equation is $4a = 79.31 - (799.35 - 792.72)$ or $a = 18.17$ cal., where a is the "true" heat of addition. He concluded that this result is unfavorable for the centric formula, "for it means that though the influence of the ring structure on the addition heat has been accounted for, this value must yet differ very considerably from that found for addition to open chain structures ($\pm$ 32 cal.)." On the contrary, using Kekulé's formula, he finds $C_{calc.} = 841.17$ (Thomsen) and $a = 30.2$ cal. In this case the true addition heat for benzene agrees closely with the ordinary value. Tombrock thus considers that the superiority of Kekulé's formula is established and also the soundness of his original premises—in adding hydrogen to closed-chain compounds part of the heat of addition is retained in the molecule; this retained energy is liberated when the chain is opened; for the ring opening extra energy is necessary. He appends (84):

Returning to our initial experimental values, it becomes evident, as has already been observed by Stohmann, that the transition from benzene into dihydrobenzene, requires a comparatively great amount of energy. Now, on further investigation in this matter this has become plausible to me if besides the angular tension of von Baeyer also the possibility of a distance tension is accepted. I take the distance between two single bonded carbon atoms to be less than that between two double united ones.

Redgrove

H. Stanley Redgrove points out that Tombrock's deduction has no argumentative value whatever for Kekulé's formula—it is a viscious circle (60). Tombrock simplified the long deduction of his previous paper (85) by reducing the question into one equation where heats of combustion are represented by chemical symbols: $C_6H_{14} - C_6H_6 = 4 H_2 \ldots + r$ cal., where r is the energy necessary to open the benzene ring. If benzene might thermochemically be considered an aliphatic substance, the value of r should be $\pm$ 0. However, $r = 991.2 - 784.1 - 4 \times 37.7 = 56.3$ cal., and this energetic difference must be ascribed to the aromatic (closed ring) character of benzene. This energy ($r = 56.3$ cal.) may be required for ring breaking. On this assumption Kekulé's formula became tenable. However, Redgrove had already shown that this supposition implies the assumption of Kekulé's structure, for the assumption that ring closing is the only difference between aliphatic and aromatic substance means, for this case, that one considers benzene merely a ring closed hexatriene. On the other hand, explaining this energy ($r = 56.3$ cal.) as necessary for four single bonds (4 times $14.7 = 58.8$), as done by Thomsen (*83*), implies the assumption that the thermal values found for aliphatic substances are equally applicable to the aromatic series, or at least that the difference in character between aliphatic and aromatic substances causes no appreciable difference in their thermochemical behavior. The latter assumption seemed at first more acceptable but implied that other thermochemical data must also be explained on mere aliphatic values. Thomsen concludes that the heat of combustion of benzene may be accounted for by Kekulé's formula on the assumption that the thermal influence of the aromatic character is considerable, or by a formula with nine single bonds on the assumption that the thermal influence of the aromatic character is negligible.

Redgrove disagreed with Tombrock's belief that Kekulé's formula could be harmonized with the thermochemical behavior of benzene (*61*). If the difference between aliphatic and aromatic substances causes, as Tombrock suggested, an appreciable difference in their thermochemical behavior, this difference must be represented as a difference in the structural formulas of the two classes of substances, but this is just what Kekulé's formula does not do. Redgrove analyzes the thermochemical behavior of benzene when repeatedly hydrogenated, in which he employs the "fundamental constants" he derived for dealing with thermochemical problems and whose values he calculated from Thomsen's experimental data. He believed his constants were more reliable than Thomsen's "because their values were not obtained by means of unlikely hypotheses (*see* [his] On the Calculation of Thermochemical Constants, 1909)."

Strain Theory

The theoretical values marked (K) are those he calculated, on the assumption that Kekulé's formula was correct; those marked (S) are calculated on the assumption that the benzene molecule contains nine single carbon-to-carbon linkages in a condition free from strain. For the reaction $C_6H_6 + H_2 = C_6H_8 + w$ cal. he states, "(K) One C:C link is replaced by one C.C link, and two H atoms are added. Therefore $w = 2H + L_1 - L_2 = \gamma - \beta = 46.0$ cal. $- 15.0$ cal. $= 31.0$ cal." Using the same theoretical values he calculates (S) $= -17.0$ cal. Stohmann's experimental value for $w = 0.8$ cal. For $C_6H_6 + 2H_2 = C_6H_{10} + x$ cal. (where x experimental is 25.8 cal.), he calculates (K) $= 62.0$ cal. and (S) $= 14.0$ cal. For the addition of $3H_2$ (experimental value $= 53.6$ cal.) he calculates (K) $= 93.0$ cal. and (S) $= 45.0$ cal. For the addition of

4 H_2 (experimental value = 64.6 cal.) he calculates (K) = 108.0 cal., and (S) = 60.0 cal. He concludes that:

The enormous differences between the experimental values and those calculated on the assumption that Kekulé's formula is correct, render this formula quite untenable. . . . The fact, however, that somewhat more heat is obtained than would be the case, according to theory, if benzene were a perfectly saturated body indicates that the benzene molecule is in a slightly strained condition.

He believed the heat effect of the strain found by subtracting the (S) value for the addition of four hydrogens from the experimental value, namely 4.6 cal., was the most reliable. By interpolation on a graph, which had already appeared (60), showing the connection between the angle of deviation of the carbon valencies and the thermochemical effect of the strain, he found that even supposing the average value of the heat effect of the strain (calculated from the above data = 10.7 cal.) were correct, this corresponds to an angle of deviation of less than 3°. Both Claus and Ladenburg's formulas entail much greater deviations than this. Redgrove therefore concluded that Baeyer and Armstrong's centric formula, which assumes that a condition of things quite different from that which obtains in the case of aliphatic compounds holds true in the benzene ring, is the preferable formula:

It may, indeed, be said that the centric formula is merely a confession of our ignorance as to the intramolecular condition of benzene, but it is better to confess our ignorance than to assert, as Kekulé's formula does, that benzene is a highly unsaturated body, in a condition of great intramolecular strain, when all of the evidence shows that it is nothing of the sort.

Ladenburg himself admitted in his 1900 history of chemistry that Kekulé's hexagon has been retained because it is superior to other formulas in many respects. It is noteworthy that because Ladenburg attached no structural significance to his formula, he was able to consider both his and other formulas throughout his work. In fact, in his 1876 pamphlet he sought to bring his and Kekulé's formulas into general agreement. Throughout his work on benzene, Ladenburg maintained an objective attitude, despite his advocacy of the prism formula. For example, in his work on the constitution of mesitylene in 1875 he proved that the three replaceable hydrogen atoms in this hydrocarbon were all of exactly equal

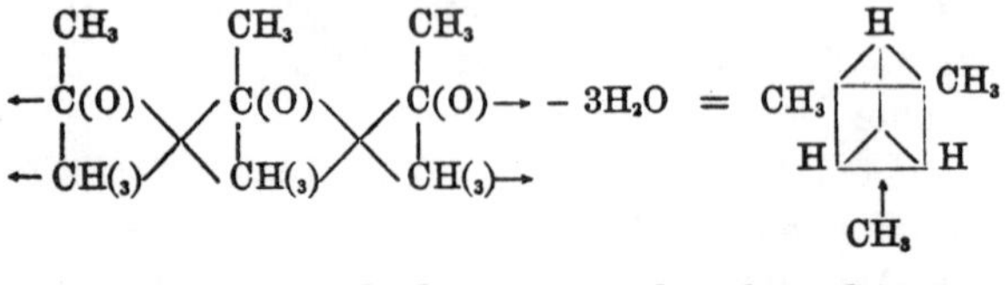

Figure 20. Ladenburg's mesitylene hypothesis

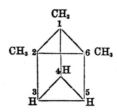

Figure 21.   Correct formula for mesitylene

value, and consequently that mesitylene was symmetrical trimethylbenzene (*46*). According to Ladenburg's hypothesis the formation of mesitylene from three molecules of acetone occurred as shown in Figure 20.
In this case the third hydrogen atom is in a different state of combination
from the other two. This formula was inconsistent with experimental
results, and the symbol for mesitylene must be written, as he indicated
in the paper, as that shown in Figure 21. However, in this case isophthalic
acid would contain two neighboring carboxyl groups, and phthalic acid
must have its carboxyl groups attached to the 1 and 3 carbon atoms.
This arrangement would also hold true for naphthalene. Such a figure is
possible but could scarcely be used as a symbolic representation of
naphthalene. Ladenburg frankly recognized that this constituted strong
evidence against the prism formula and concluded that "there is at present time no symbolic representation of benzene which satisfies all
requirements."

Baeyer

The most complete refutation of Ladenburg's formula was furnished
by Baeyer, who started his research on the reduced phthalic acids in
1886. He pointed out that although benzene derivatives were obtainable
from hexamethylene compounds, it does not follow that only hexamethylene compounds must result when benzene compounds are reduced. He
admitted the possibility of the formulas of Kekulé, Claus, Dewar, and
Ladenburg, although in the last case ditrimethylene derivatives should
be possible reduction products, being formed by severing two of the
prism edges. He attempted to solve the problem by systematically studying the reduced phthalic acids.

Ladenburg's prism formula accounts for one monosubstitution derivative and three diderivatives. Moreover, it agrees with certain simple
syntheses of benzene derivatives—e.g., acetylene and acetone. However,
according to Baeyer (*3*), it fails to explain the formation of dioxyterephthalic ester from succinosuccinic ester (Figure 22) unless one assumes
that the transformation of these substances is attended by a migration of
the substituent groups for succinosuccinic ester has either the formula I

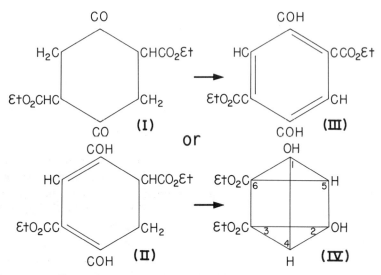

Figure 22. Formation of dioxyterephthalic ester

or II. Oxidation of the free acid gives dioxyterephthalic acid, in which the para positions must remain substituted as in I and II. By projecting Ladenburg's prism on a plane and numbering the atoms to correspond with Kekulé's formula (—i.e., the 1:2 and 1:6 should be ortho positions, 1:3 and 1:5 meta and 1:4 para), and following out the transformation on the Ladenburg formula, then an ortho-dioxyterephthalic acid, IV, should result—a fact denied by experience and inexplicable unless a wandering of atoms is assumed. Kekulé's formula, III, is, on the other hand, according to Baeyer, in full agreement. This explanation was seriously challenged by Ladenburg (50) and by A. K. Miller (57). The transformation is not one of the oxidation of a hexamethylene compound to a benzenoid compound since only two hydrogen atoms are removed. Succinosuccinic ester behaves both as a ketone and as a phenol, thereby exhibiting what was then called "desmotropy." Assuming that the ketone formula indicates the constitution, then in Baeyer's equation we have a migration of a hydrogen atom whereas to bring Ladenburg's formula into line, an oxygen atom must migrate.

Again the refutation of Ladenburg's formula was furnished by Baeyer (5). He showed that, of the three hexahydrophthalic acids, only the ortho compound readily forms an anhydride; the meta compound must be heated with acetyl chloride, and the para compound under no conditions forms an anhydride. According to Ladenburg's formula, phthalic acid on reduction should produce a meta- or para-cyclohexane derivative, depending on which set of bonds is removed. The process may be graphically represented (Figure 23) in such a way that after

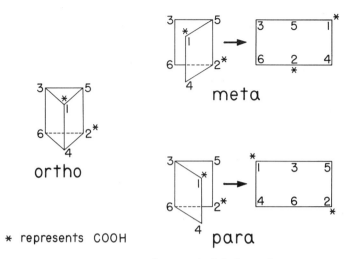

ortho

para

meta

* represents COOH

Figure 23.   Reduction of phthalic acid

the three links are removed, the two prism faces are folded back like the covers of a half-opened book. It is difficult to reconcile these formulas with the existence of an anhydride which is more stable than that produced by carboxyls in the ortho position as denoted by the Kekulé formula.

Spatial Considerations

Although Kekulé's formula accounted for facts known then and suggested many new lines of work, his hypothesis represented benzene as existing in two-dimensional space and therefore could be regarded as only symbolic. It could not represent actual conditions in the benzene molecule. This paper will not discuss the arguments for and against each of the space formulas; but it may be observed that for any formula to be satisfactory, it must represent in a simple fashion the symmetry of the molecule, the process of hydrogenation, the anhydride formation of ortho compounds, and the relation to naphthalene and other polynuclear hydrocarbons. These points have all been discussed in detail by Graebe (23), who concluded that Kekulé's formula was the only one to meet the many demands made upon it. With the single exception of Kekulé's formula, there is one inherent defect in all. Unless the carbon and hydrogen atoms lie in the same plane, replacing two hydrogen atoms by different groups in the ortho or meta positions leads at once to asymmetry and to the possibility of optical enantiomorphs. It has already been shown that neither among artificial nor natural products have substituted benzene derivatives of this character been observed.

Thus, the essential features of Kekulé's structure—a closed chain of six carbons with one hydrogen on each—were rapidly accepted; only Ladenburg seriously challenged them. The problem of the fine structure, however, which centers on the disposition of the fourth valence had proved a most fruitful field for imaginative speculation almost until the present day. Among a host of suggestions, most of which used special symbols to depict the particular and peculiar nature of benzene (referred to as "ignorance symbols"), those of Kekulé and Thiele are the most noteworthy (71).

It is difficult to overestimate the importance of Kekulé's structure on the subsequent development of carbon chemistry. His hypothesis not only accounted for the facts then known concerning the chemistry of aromatic compounds but suggested many new lines of work. As a result it became the dominant thought in organic chemistry until the significance of three-dimensional space was recognized. Kekulé's generalization enabled chemists to arrange their facts systematically and consider them intelligently. Since the accuracy of Kekulé's predictions also inspired a belief in the utility of a legitimate hypothesis in chemistry, it did more to elevate the deductive side of the science than almost any other investigation up to that time. It is worth repeating that "Kekulé's work stands preeminent as an example of the power of ideas."

Renewed Interest in the Prism Formula

Ladenburg's criticism which caused Kekulé's hypothesis to be thoroughly scrutinized and even modified, as in the case of Kekulé's 1872 oscillation formula, would have been sufficiently important to ensure the prism formula a place in history. That this formula may have a greater significance has been brought out in the 1964 work of Viehe and his colleagues (91, 92). First a brief review of 20th century developments is in order.

The subject of the prism formula was never completely closed. An unsuccessful attempt to synthesize a prism structure was made by Farmer in 1923 (22). Also in 1923 Shearer noted that the crystal structure of benzene requires a fourfold symmetry and is best represented by the Ladenburg or Dewar formulas (66). Nevertheless, by 1900 the prism formula presented no real challenge to the Kekulé hexagon. However, interest in the prism formula continues.

Significant experimental and theoretical work on the structure of benzene was done in the 1930's. Physical measurements confirmed that the benzene molecule has a planar symmetrical structure (68). Kekulé's oscillation formulas were reinterpreted as a hybrid formed by linear combination of the two cyclohexatriene forms according to the valence bond method. Kekulé's "oscillation" became "resonance"; quantum mechanics was applied to valence theory. Resonance energies of many aromatic compounds were determined from heats of combustion or hydrogenation and correlated with quantitative valence bond calculations, but the valence bond method did not account for the unique role of six electrons in aromatic molecules. Molecular orbital theory, particularly as developed by Hückel, provided the explanation needed (26, 27). Consideration of the molecule as a whole replaced localized bond descriptions by designating the shapes, positions, and energies of the orbitals which the electrons occupy. Conclusions were based on mathematical and experimental foundations. Molecular orbital theory predicted allowed numbers of electrons for stable closed shells in molecules. Only the double-bond electrons were placed in molecular orbitals (π-orbitals) in Hückel's theory. Filling any shell in a monocyclic conjugated system requires $4n+2$ π-electrons, two for the lowest orbital, and four more for each occupied degenerate pair. Thus, benzene with six π-electrons has a closed shell and is stable, and the modern pictorial

*Figure 24. π-bonding
in benzene*

representation of benzene is a single structure. A single plane of carbon atoms is held together by single bonds (sigma bonds) composed of 12 electrons. In addition six π-electrons which form dumbbell-shaped areas of high electron density above and below the plane of carbon atoms, are involved in the bonding. The overlap of the π-electrons (Figure 24) constitutes π-bonding and increases the stability of the ring.

Valence Bond Theory

While the structures proposed by Dewar, Ladenburg, and others, conceived as planar systems, had no significance as physical, isolable entities independent of benzene itself, the Dewar formula persisted as a minor electronic contributor in the valence bond method. If the "valence-bond isomeric formulae" are regarded as being nonplanar, however, it is implied that they are individual compounds different from benzene and capable of existing in their own right. The isomerism of these molecules is then caused only by the different arrangements of their valences (94). The steric strain in the "bent" or nonplanar valence isomers of benzene should result in unfavorable energies relative to the resonance-stabilized planar Kekulé benzene. Accordingly it was felt that they could not be isolated easily since they should isomerize readily to the planar benzene.

In 1962, however, E. van Tamelen and S. Pappas produced the first Dewar benzene structure 1,2,4,-tri-*tert*-butylbicyclohexadiene by subjecting 1,2,4-tri-*tert*-butylbenzene to ultraviolet light (*87*). A year later, the same authors synthesized Dewar benzene itself (*88*). Ultraviolet irradiation was used to transform *cis*-1,2-dihydrophthalic anhydride to a cyclobutene derivative. Treating this intermediate, bicyclo-(2.2.0) 5-hexen-2,3-dicarboxylic acid anhydride, with lead tetraacetate produced the Dewar benzene. It's half-life is about two days at room temperature. Two further syntheses of Dewar benzene have recently been reported (*2, 19*), but of still greater interest in connection with the Ladenburg formula is some recent Belgian work.

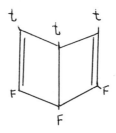

*Figure 25.   Dewar benzene
derivative (t=tert-butyl)*

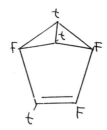

Figure 26.   Benzvalene form

H. G. Viehe and his co-workers entered the field of the valence isomers of benzene by way of their work on heterosubstituted acetylenes. They noted that *tert*-butylfluoroacetylene undergoes spontaneous oligomerization to trimers and tetramers (*91, 92*). Two valence-bond isomers of benzene can be isolated from the trimer fraction which makes up two-thirds of the reaction product. The trimer fraction consists of roughly equal parts of a Dewar benzene derivative (Figure 25) and the compound they named "benzvalene" (Figure 26). Of considerable interest to our discussion was the 1–3% of a product which they first formulated on the basis of preliminary molecular weight determinations as Ladenburg benzene, or as they named it "prismane" (*91, 92*). The lack of all aromatic character further convinced them that they had indeed isolated the Ladenburg prismatic structure. Later x-ray data and redetermination of the molecular weight when more material had been isolated showed that the compound is a tetramer of *tert*-butylfluoroacetylene and not Ladenburg benzene (*93*).

While a synthesis of the Ladenburg prism has not yet been achieved, van Tamelen has expressed the view that he would be surprised indeed if prismane or "Ladenburg benzene" were not made within the next few years (*89*). Breslow's recent review of aromatic character amply illustrates that "one hundred years after Kekulé, chemists are still trying to understand aromatic character" (*7*).

A study of the history of benzene formulas revealed that the Ladenburg formula was for a long time the strongest contender as an alternative to the Kekulé formula. Recent laboratory studies confirm that the synthesis of "Ladenburg benzene" is still of current interest.

Literature Cited

(1) Armstrong, Henry E., *J. Chem. Soc.* **51**, 258 (1887).
(2) Arnett, E. M., Bollinger, J. M., *Tetrahedron Letters* **1964**, 3803.
(3) Baeyer, A., *Ber.* **19**, 1797 (1886).
(4) Baeyer, A., *Ann.* **245**, 193 (1888).
(5) Baeyer, A., *Ann.* **258**, 145 (1889).
(6) Barth, L., *Monatsh.* **1**, 869 (1880).
(7) Breslow, Ronald, *Monatsh.* **43**, 90 (1965).
(8) Brühl, J. W., *Ann.* **200**, 139 (1880); **203**, 255 (1880).
(9) Brühl, J. W., *Monatsh.* **2**, 716 (1881).
(10) Brühl, J. W., *J. Prakt. Chem.* **35**, 181 (1887).
(11) Brühl, J. W., *Ber.* **20**, 2288 (1887).
(12) Brühl, J. W., *J. Prakt. Chem.* **49**, 201 (1894).
(13) Brühl, J. W., *Ber.* **27**, 1065 (1894).
(14) Claus, Adolph, *J. Prakt. Chem.* **37**, 455 (1888).
(15) Claus, A., *J. Prakt. Chem.* **49**, 505 (1894).
(16) Claus, A., "Theoretische Betrachtungen und deren Anwendung zur Systematik der Organischen Chemie," p. 208, O. Hollander, Freiberg, 1867.
(17) Collie, J. N., *J. Chem. Soc.* **71**, 1013 (1897); **109**, 561 (1916).
(18) Couper, Archibald S., *Compt. Rend.* **46**, 1157 (1858). English translation in "Classics in the Theory of Chemical Combination," O. T. Benfey, ed., pp. 132-135, Dover, New York, 1963.
(19) Criegee, R., Zanker, F., *Angew. Chem.* **76**, 716 (1964) (*Angew. Chem. Intern. Ed. Engl.* **3**, 695 (1964)).
(20) Dewar, James, *Proc. Roy. Soc. Edinburg* **6**, 81 (1869).
(21) Erlenmeyer, Emil, *Ann.* **316**, 57 (1901).
(22) Farmer, E. H., *J. Chem. Soc.* **123**, 3332 (1923).
(23) Graebe, Carl, *Ber.* **35**, 526 (1902).
(24) Horstmann, A., *Ber.* **21**, 2211 (1888).
(25) Hübner, H., Petermann, A., *Ann.* **149**, 129 (1869).
(26) Hückel, E., *Z. Phys.* **70**, 204 (1931).

(27) Hückel, E., Z. Elektrochem. **45**, 752, 760 (1937).
(28) Kekulé, F. A., Bull. Soc. Chem. **1**, 98 (1865).
(29) Kekulé, F. A., Bull. Acad. Roy. Belg. **19**, 551 (1865).
(30) Kekulé, F. A., "Lehrbuch der Organischen Chemie," vol. 2, Enk, Erlangen, 1866.
(31) Kekulé, F. A., Ann. **137**, 129 (1866).
(32) Ibid., p. 160.
(33) Kekulé, F. A., Z. Chem. **3**, 216 (1867).
(34) Kekulé, F. A., Ber. **23**, 1302 (1890). For English translations see Japp, F. R., J. Chem. Soc. **73**, 97 (1898) and Benfey, O. T., J. Chem. Ed. **35**, 21 (1958).
(35) Kekulé, F. A., Ann. **162**, 77, 309 (1872).
(36) Kekulé, F. A., Ann. **221**, 230 (1883).
(37) Kekulé, F. A., Strecker, O., Ann. **223**, 170 (1884).
(38) König, B., Chem. Z. **29**, 30 (1905).
(39) Koerner, W., Giorn. Sci. Nat. Ed. Acon. **5**, 241 (1869).
(40) Koerner, W., Gazz. Chim. Ital. **4**, 305 (1874).
(41) Ladenburg, Albert, Ber. **2**, 140 (1869).
(42) Ladenburg, A., Ber. **2**, 272 (1869).
(43) Ibid., p. 273.
(44) Ladenburg, A., Ber. **5**, 322 (1872).
(45) Ladenburg, A., Ber. **7**, 1684 (1874).
(46) Ladenburg, A., Ann. **179**, 163 (1875).
(47) Ladenburg, A., Ber. **8**, 1214 (1875).
(48) Ladenburg, A., Ber. **9**, 1524 (1876); **10**, 1123 (1877).
(49) Ladenburg, A., Ber. **10**, 1154 (1877).
(50) Ladenburg, A., Ber. **19**, 971 (1886); **20**, 62 (1886).
(51) Ladenburg, A., "Theorie der Aromatischen Verbindungen," p. 4, F. Vieweg, Braunschweig, 1876.
(52) Ibid., p. 27.
(53) Ibid., p. 23.
(54) Marckwald, W., Ann. **279**, 1 (1894).
(55) Marckwald, W., Ber. **35**, 703 (1902).
(56) Meyer, L., "Modern Theories of Chemistry," pp. 239-40, Bedson and Williams, London, 1888.
(57) Miller, A. K., J. Chem. Soc. **51**, 208 (1887).
(58) Muir, M. M. P., "A History of Chemical Theories and Laws," 1st ed., p. 298, Macmillan, New York, 1909.
(59) Noelting, E., Ber. **37**, 1015 (1904).
(60) Redgrove, H. S., Chem. News **106**, 173 (1912).
(61) Redgrove, H. S., Chem. News **106**, 224 (1912).
(62) Rosenstiehl, A., Bull. Soc. Chim. **11**, 385 (1869).
(63) Sachse, H., Ber. **21**, 2530 (1888).
(64) Schiff, R., Ann. **220**, 303 (1883).
(65) Schröder, H., Ann. **15**, 636 (1882).
(66) Shearer, G., Proc. Phys. Soc. London **35**, 81 (1923).
(67) Smile, Samuel, "The Relations Between Chemical Constitution and Some Physical Properties, Longmans, Green, and Co., London, 1910.
(68) Staab, H. A., "Einfuhrung in die Theoretische Organische Chemie," p. 80, Verlag Chemie, Berlin, 1880.
(69) Stohmann, F., J. Prakt. Chem. **19**, 115 (1879).
(70) Stohmann, F., Sachs. Ges. Wiss. 1893, 477.
(71) Thiele, J., Ann. **306**, 125 (1899).
(72) Thomsen, H. P. J. J., Ann. **205**, 133 (1880).
(73) Thomsen, H. P. J. J., Ber. **13**, 1388 (1880).
(74) Thomsen, H. P. J. J., Ber. **13**, 2166 (1880).

(75) *Ibid.*, p. 2167.
(76) Thomsen, H. P. J. J., *Ber.* **13,** 1806 (1880).
(77) Thomsen, H. P. J. J., *Ber.* **13,** 1808, 1811 (1880).
(78) Thomsen, H. P. J. J., *J. Prakt. Chem.* **23,** 157 (1881).
(79) Thomsen, H. P. J. J., *Ber.* **15,** 328 (1882).
(80) Thomsen, H. P. J. J., *Ber.* **15,** 66 (1882).
(81) *Ibid.*, p. 69.
(82) Thomsen, H. P. J. J., *Ber.* **19,** 2944 (1886).
(83) Thomsen, H. P. J. J., "Thermochemistry," translated by A. Burke, p. 394,
      Longmans, Green and Co., London, 1906.
(84) Tombrock, Willebrord, *Chem. News* **106,** 155 (1912).
(85) Tombrock, W., *Chem. News* **106,** 201 (1912).
(86) van't Hoff, J. H., *Ber.* **9,** 1881 (1876).
(87) van Tamelen, E. E., Pappas, S. P., *J. Am. Chem. Soc.* **84,** 3789 (1962).
(88) van Tamelen, E. E., Pappas, S. P., *J. Am. Chem. Soc.* **85,** 3297 (1963).
(89) van Tamelen, E. E., *Chemistry,* **38,** 6 (1965).
(90) Vaubel, W., *J. Prakt. Chem.* **49,** 308 (1894); **50,** 362 (1894).
(91) Viehe, H. G. *et al., Angew. Chem.* **76,** 922 (1964) (*Angew. Chem.
      Intern. Ed. Engl.* **3,** 746, 755 (1964)).
(92) Viehe, H. G. *et al., Chem. Eng. News* **42,** 38 (Dec. 7, 1964).
(93) Viehe, H. G. *et al., Angew. Chem.* **77,** 770 (1965) (*Angew. Chem.
      Intern. Ed. Engl.* **4,** 746 (1965)).
(94) Vogel, E., *Angew. Chem.* **74,** 829 (1962) (*Angew. Chem. Intern. Ed.
      Engl.* **2,** 1 (1963)).
(95) Walker, O. J., *Ann. Sci.* **1939,** 42.
(96) Watts, E. A., "Dictionary of Chemistry," suppl. vol., Art. Aromatic Series.
(97) Wroblewsky, E., *Ann.* **168,** 147 (1873); **192,** 196 (1878).

RECEIVED January 3, 1966.

Courtesy Verlag Chemie

Low relief on base of Kekulé statue at Bonn.

INDEX

195